Analysis of Machining and Machine Tools

Steven Y. Liang • Albert J. Shih

Analysis of Machining and Machine Tools

Steven Y. Liang
George W. Woodruff School of
Mechanical Engineering
Georgia Institute of Technology
Atlanta, Georgia
USA

Albert J. Shih
Mechanical Engineering
University of Michigan
Ann Arbor, Michigan
USA

ISBN 978-1-4899-7643-7
ISBN 978-1-4899-7645-1 (eBook)
DOI 10.1007/978-1-4899-7645-1

Library of Congress Control Number: 2015953017

Springer New York Heidelberg Dordrecht London

Printed on acid-free paper

Springer US is part of Springer Science+Business Media (www.springer.com)

Preface

Among many viable manufacturing processes to date, machining is no doubt one of the most important, largely inasmuch as machining's unpatrolled capability to shape products and/or control their qualities and properties with precision, efficiency, and cost-effectiveness. Without a solid command of machining science and engineering, many parts and components with stringent geometry and functionality requirements cannot be satisfactorily produced to satisfy today's demanding markets in aerospace, automotive, energy, biomedical, consumer electronics, and many others.

The effective implementation of machining process hinges upon a wide range of technical issues, including design, selection, and optimization of machine, tooling, operation configuration, process parameters, and their entangling relationships. A holistic and in-depth understanding of these issues is essential to pursue the eventual efficiency and quality of machining. The technology behind these issues is not just art and empiricism, but, more importantly, science. Only through scientific reasoning and physical understanding can the convoluted interactions and intricate details of these issues be coped with successfully in the interest of product design, process planning, as well as machine and tooling optimization.

There is no doubt that the scientific theories and methodical principles are the essential pillars to support effective and competitive machining processes. Based on this understanding in mind, this textbook is written with a mission to deliver the fundamental science and mechanics of machining and machine tools. Instead of resorting to a large amount of experimental test data, empirical observations, and ad hoc rules of thumb, this book has made a conscientious effort to present the systematic and quantitative knowledge in the form of process mechanics and physics. Information in the form of materials' properties, machining data, and parameter recommendations, are commonly available online and in commercial catalogs nowadays; however, the authors believe that knowledge is a step higher than information and it can only come as a result of rigorous scientific analysis and reasoning. The acquired knowledge, not just information, is actually what it takes to effectively handle situations with unknowns—new materials, new tools, new machines, and new configurations. The primary objective of this book is to thus provide a platform for learning, exercising, and hopefully mastering the basic theories and fundamental principles so that effective implementation of machining engineering is possible and further development of scientific understanding on machining and machine tools is stimulated.

One key feature of this book is that it has been populated with examples and exercise problems in every chapter. The examples and problems are given in the context of real-life industrial applications to offer practical engineering relevance. Meanwhile, scientific principles and theoretical fundamentals are blended into the solution methods to stress on analytical reasoning.

The book begins with the introduction of technology and market demands in Chap. 1. It is followed by the simplest case of single-point turning process in Chap. 2 and a more complicated case of multiple-point machining in Chap. 3. Chapter 4 is dedicated to stochastically distributed multiple-point grinding. Following from the discussion on process characteristics, the machine tool equipment characteristics are elaborated in terms of the basic components in Chap. 5 and the accuracy metrology in Chap. 6. The machining mechanics is introduced in Chap. 7, and there is an extended discussion on shear stress in Chap. 8 to explain the notation of specific machining energy variation. Apart from the mechanical aspect, Chap. 9 discusses the thermal aspect of heat generation and the resulting cutting temperature distributions. Chapter 10 overviews the dynamic stability of machining process and its relevance to various modes of chatter vibrations. Chapters 2–10 are on traditional machining, while Chaps. 11–14 discuss nontraditional machining. Chapter 11 covers electrical discharge machining, Chap. 12 encompasses various machining involving chemical dissolution, and Chap. 13 elaborates on high-energy beam machining. Chap. 14 discusses the machining processes with biomedical materials as the workpiece.

University senior students, graduate students, manufacturing engineers, and research center staff members in industry are the targeted readers. As the book stresses on the use of broad-based engineering mechanics for physical understanding and problem solving, prior background on mechanics of materials, statics, dynamics, control theory, and heat transfer would be recommended for in-depth appreciation of the book content. At a graduate level, the book can also serve as a support for thesis and dissertation studies that call for scientific principles and application relevance.

The content of this book is an accumulation of many years of teaching experience between Professor Steven Y. Liang of Georgia Institute of Technology and Professor Albert J. Shih of University of Michigan. The development of this book is indebted to many former graduate and undergraduate students who took courses based on the book manuscript and helped suggest and modify many parts of it. The authors would like to acknowledge, in particular, the contributions of Dr. Yong Huang (University of Florida, USA), Dr. Kuan M. Li (National Taiwan University, Taiwan, ROC), Dr. Bruce L. Tai (Texas A&M, USA), Dr. Roland Chen (Washington State University, USA), Mr. Lei Chen (University of Michigan, USA), and Mr. Yao Liu (Donghua University, PRC) in proofreading and editing many parts of this text.

Acknowledgment

The professional support from the Springer publication team, particularly Marta Moldvai and Pramod Prasad, are greatly appreciated. The authors also acknowledge the research support in machining and machine tools from the United States National Science Foundation (NSF), National Institute of Standards and Technology (NIST), Ford Motor Company, Boeing, General Motors, Caterpillar, Cummins, and many others. The support of our research teams and practical problems in the machining line in production have been essential to develop the knowledge and build the foundation of this book.

Contents

About the authors

Steven Y. Liang is Morris M. Bryan, Jr. Professor in Mechanical Engineering for Advanced Manufacturing Systems at Georgia Institute of Technology, USA. He was the institute's founding director of Precision Machining Research Consortium (PMRC) and director of the Manufacturing Education Program (MEP) from 1996 to 2008. Dr. Liang has conducted research in various fields of precision machine tools, machining mechanics, and intelligent manufacturing. He has authored in excess of 350 book chapters and archival journals and conference papers, and has edited several conference proceedings in the general areas of manufacturing science and technology. He has been invited to deliver more than 70 keynote speeches and seminars at manufacturing industries, peer institutions, and professional conferences in over 20 countries on various topics related to manufacturing.

Dr. Liang served as the president of the North American Manufacturing Research Institution of the Society of Manufacturing Engineers (NAMRI/SME) and the chair of the Manufacturing Engineering Division of The American Society of Mechanical Engineers (MED-ASME). He is the recipient of many awards, including the SME Robert B. Douglas Outstanding Young Manufacturing Engineer Award, Society of Automotive Engineers Ralph R. Teetor Educational Award, ASME Blackall Machine Tool and Gage Award, and Outstanding Alumni Award of National Cheng Kung University (NCKU) in Taiwan. Dr. Liang is a fellow of both ASME and SME. He received his B.S. from NCKU, M.S. from Michigan State University, and Ph.D. from University of California at Berkeley.

Albert J. Shih is a professor of mechanical engineering, biomedical engineering, and integrative systems + design (ISD) at the University of Michigan, USA. He is the associate chair of ISD (2013–present) and director of the Global Automotive and Manufacturing Engineering Program (2011–present). He was a manufacturing process development engineer, specialized in advanced grinding, from 1991 to 1998, at Cummins Inc. Dr. Shih's research area is design and manufacturing. He is a pioneer in biomedical manufacturing, the application of manufacturing technologies to advance the safety, quality, efficiency, and speed of health-care service, and biomedical science. He has authored in excess of 300 book chapters and archival journal and conference papers in manufacturing and biomedical sciences.

Dr. Shih has served as the chair of Scientific Committee of the North American Manufacturing Research Institution of the Society of Manufacturing Engineers (NAMRI/SME). He is the recipient of many awards, including the Fulbright Scholar, ASME Milton Shaw Manufacturing Research Medal, Society of Automotive Engineers Ralph R. Teetor Educational Award, Mechanical Engineering Alumni Award of National Cheng Kung University, and Best Paper awards in ASME International Manufacturing Science and Engineering Conference (MSEC), North American Manufacturing Research Conference (NAMRC), International Conference on Frontiers of Design and Manufacturing (ICFDM), and several other manufacturing and medical conferences. Dr. Shih is a fellow of ASME and SME and associate member of CIRP. He received his B.S. and M.S. from National Cheng Kung University and Ph.D. from Purdue University.

Introduction

1

1.1 Basic Definitions

Machining is the process to shape any raw material (workpiece), metal or non-metal, into a part (product) or to improve the tolerance and surface finish of a previously formed workpiece by removing a portion of the raw material. This can be done either mechanically (turning, drilling, milling, grinding, water-jet machining, ultrasonic machining, etc.), chemically (chemical machining, electrochemical machining, etc.), electrically (electrical discharge machining), or thermally (laser machining, electron beam machining, etc.).

Metal cutting is a process in which excess material is removed by a harder tool, through a mechanical process of extensive plastic deformation or controlled fracture. Metal cutting, in this definition, is a subset of machining.

In a broad sense, DIN 69651 defines machine tools as "machines provided with a power source, for the main part non-portable, which are used for a variety of production procedures, with the aid of physical, chemical or other processes. … Machine tools bring the interacting tool and work together in such a way that after defined relative motions between them, a geometrically definable work form (finished component) results at the end of the production process." According to this definition, a machine tool is a manufacturing tool that comes in the form of a machine. It can perform one or a combination of processes such as cutting, casting, forming, or joining. On the other hand, the term machine tool is commonly used to describe the hardware system (equipment) that performs machining only. In this narrower context, machine tool refers to the tool that is used to fabricate machines.

Machining is the backbone technology for a large number of manufacturing systems. It may be used in either the primary manufacturing process or form an important part of preparing the tooling for other processes like forming or molding. Machining and machine tools are important over other manufacturing processes because:

S. Y. Liang, A. J. Shih, *Analysis of Machining and Machine Tools*,
DOI 10.1007/978-1-4899-7645-1_1

- Machining is extremely precise: It can create geometric configurations, tolerances, and surface finishes often unobtainable by any other technique (such as casting, forming, or joining). For example, generally achievable arithmetic average surface roughness is 10–20 μm in sand casting, 2–5 μm in die casting, 5–10 μm in forging, and 0.5–1 μm in turning. In precision machining (superfinishing, lapping, diamond turning, etc.) it can be 0.01 μm or better. The achievable dimensional accuracy in casting is 0.8–2 %, depending on the thermal expansion coefficient; in metal forming it is 0.05–0.3 %, depending on the yield strength and stiffness; in machining it can be infinitely good since the dimensional accuracy becomes independent of the size of the workpiece.
- Machining is highly flexible: In machining, the shape of the final product is programmed; therefore many different parts can be made on the same machine tool and just about any arbitrary shape can be machined. The fact that the product contour is created by the path rather than the shape of the cutter makes the process extremely flexible, agile, and economical for prototyping and small batch manufacturing. The cutting tools can be mass produced in standardized shapes. On the contrary, casting, molding, and forming processes require one dedicated tool for each product, which makes them much less flexible.

Machine tool consumption was about $9.2 billion in the USA and $58 billion worldwide in 2014. About 30–40 % of the US machine tool market is automotive-related, including direct and indirect operations such as die making and bearing manufacturing. An increase in machine tool consumption is expected in the next decade because of the lack of skilled workers, stronger need of automation, increasing demand for precision, and stricter environmental regulations. There is a strong competition from overseas.

The knowledge involved in the application of machine tools, metal cutting, computer-aided manufacturing, and automatically controlled processing are broad and diversified. Although it is possible to develop engineering expertise in each area based on hands-on operations and empirical know-how, the ability to extend beyond ordinary observation and the luxury to integrate various areas to achieve complicated tasks has to rely on fundamental analysis and scientific know-why. The objective of this book is to offer a blend of hands-on experience and analytical background to provide an appreciation of the science of machining and machine tools. The background understanding developed in this course will lend itself to productivity analysis, strategic process planning, machinery design/selection, automation development, and the integration of manufacturing system.

1.2 Historical Development of Machining Technology

Industrial production first started at the beginning of the nineteenth century. It generally replaced the manufacture of goods by craftsmen. The introduction of machinery, as well as the division of labor, permitted the production of goods in large quantities. The Industrial Revolution, which found its origin in the new production methods, was made possible through, among other factors, the invention of the steam engine and then the electric motor. These permitted the mechanization of the power source of production machines. In the area of the metal-working industries, the development of machine tools began at that time. In parallel, vast advances were made in metallurgical and technological fields.

In all aspects of industrial production, machine tools play a major role as a means of producing the goods. From this, it follows that this branch of machine tool manufacture is of central economic importance. The machine tool industry provides all the production installations for the needs of the whole metal-working industry. The quality of the finished product and manufacturing costs are both dependent on the state of technical development of the machine tool industry.

In the late 1700s, after the further advancement of steam engine, machine shops using machine tools were developed using steam power. Eli Whitney (1765–1825) developed the concept of interchangeable parts for soldiers in the battlefield to share parts of their rifles. The concept enabled the mass production using a series of machine tools. Mass production opened the opportunities for low-cost production and generated needs for machines, tools, metrology, and factory management.

The scientific study of metal cutting started in 1850s. One of the standout researchers is Frederick Taylor, who worked at the Midvale Iron Works (near Philadelphia) and nearby Bethlehem Steel Works, and he developed the scientific management as well as metal-cutting theory and practice, which was critical to implement his management theories. Taylor, worked jointly with metallurgist Maunsel White, invented the high-speed steel-cutting tool material to increase the cutting speed and improve the productivity. Taylor also discovered the importance of tool temperature on tool life, which led to creation of the tool life equation.

The advancement in machine tool, tooling, and factory management made the mass production using in-line transfer machine possible. In 1900s, Henry Ford revolutionized the automotive production. The rise in automotive industry further advanced the machining, machine tool, and tooling technology.

Wars brought the revolutions of machining technology. During World War II, Germany developed the tungsten carbide cutting tool. After the war, under the support of the US Navy, the computer numerical control machine was invented. Advanced aircraft utilized advanced light-weight, high-temperature materials and created the need for ceramic cutting tools as well as tool coatings for efficient machining. Computer as well as mathematical modeling also changed machining from an art to science-based practice. At the turn of twenty-first century, researchers are continuing to push the envelope of machining technology. The micro- and nano-scale machining, machining of hard and brittle materials, semiconductor manufacturing, minimal quantity lubrication (MQL) machining, and biomedical machining are just a few examples of the continuing advancements of machining in the coming years.

1.3 Overview of Machining Processes and Systems

Traditional metal-cutting process, a subset of machining, can be categorized into following processes:

1. Turning: Performed on a lathe to produce externally cylindrical or axisymmetric parts
2. Boring: A lathe process to produce internally axisymmetric parts
3. Reaming: A process to enlarge the hole size and improve roundness and surface finish
4. Drilling: A hole-making process using a tool called drill to create or enlarge a hole
5. Milling: Material removed by a rotating cutter, usually having multiple cutting edges
6. Broaching: Using tools of gradually enlarging shape to remove material
7. Tapping and threading: Using a sharp, pointed tool to generate internal or external thread
8. Grinding: Using bonded hard abrasive as the tool for material removal
9. Honing: A process utilizing the expanding tool, called honing stones, to correct the axial and radial distortions from previous operations
10. Burnishing: Using a hard, smooth roller or ball as the tool to press against the work surface and generate the finish surface through plastic deformation
11. Deburring: A process to remove burrs, the undesired projections of material beyond edges of machined workpiece due to work material plastic deformation at the exit of the machining process

In addition to these traditional metal-cutting processes, other processes using electrical, chemical, and thermal energy for removal of the work material have been developed. These so-called nontraditional machining processes are:

- Electrical discharge machining (EDM): This process uses electrical sparks between the tool electrode and workpiece to erode and remove work materials. The electrode can be shaped form (die-sinking EDM) or continuously running wire (wire EDM).
- Electrochemical machining (ECM): This process uses electrolysis to remove the work material.
- Waterjet: High-pressure water can be used to cut soft work materials, such as bread or carpet, cleanly and quickly. In the abrasive waterjet machining, abrasive particles are mixed in the high-pressure water and are used to cut metals, concrete, granite, and other materials.
- Laser machining: The concentrated laser energy is used to melt and remove materials.
- Plasma machining: The hot plasma is used for flexible, low-cost cutting of materials.
- Ion-beam machining: The energy can input via ionization to remove the work material. It usually removes a small amount of material, such as the optical lens, after form polishing.

The combination of machining processes, commonly called operations in the factory production, to produce one component or a group of components is called the

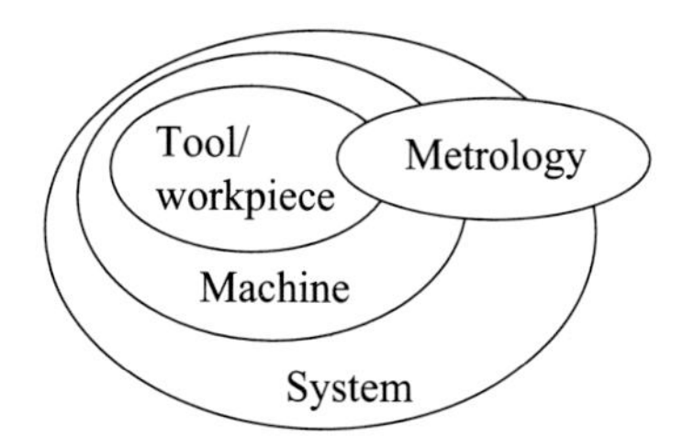

Fig. 1.1 Relationship among the tool/workpiece, machine, system, and metrology (Aachen model)

machining systems. Due to the economy of scale to reduce cost, a factory can involve a large number of machines, like hundreds or thousands of machines, in operation in one plant. The integration of machines with the measurement system, shop floor operations system, and production system forms a "manufacturing system." Such manufacturing system involves extensive capital investment and long-term return on investment. In the era of short product life cycle and quick product development time, the emphasis moved toward flexibility (flexible manufacturing systems (FMSs)) in the 1970s and 1980s and reconfigurability (reconfigurable manufacturing systems (RMSs)) in the 1990s and early twenty-first century. While studying the individual machining processes, it is important not to lose sight that every machine will be used in a factory and is part of a manufacturing or production system.

A good way to describe the manufacturing system is the so-called Aachen model, as shown in Fig. 1.1. The interaction between tool and workpiece is the heart of manufacturing processes. Most of this book investigates the physical and chemical phenomena that occur at the tool–workpiece interface. The tool and workpiece are both supported and moved by the machine to create the required machining action. The book provides an overall view of machines and machine components. Machines are part of the manufacturing system. It could be the manufacturing or production system that manages the factory and business. Metrology integrates across all three levels of tool/workpiece, machine, and manufacturing system.

Globalization in manufacturing and the shift of production to developed countries are key developments that took place in the early twenty-first century. Machining and machine tool technologies become ubiquitous and available to manufacturers around the world. Contract manufacturing to produce key components and assemble final product at strategic locations around the world and utilizing the global logistics to transport to customers have revolutionized the traditional production paradigm and enabled the global manufacturing. One of the evident changes is the change in machine tool production worldwide. Figure 1.2 shows the percentage of worldwide machine tool production from 1980 to 2014. China, Germany, Japan, Italy, South Korea, Taiwan, Switzerland, and the USA are consistently the top machine tool producers in the world. The dynamic change of machine tool industry and the rise of Asian countries are evidence of the shift in manufacturing activities. The decline of production is noted due to a worldwide recession, and it has been reverted by a significant gain in 2009 and 2012 as shown in Fig. 1.3.

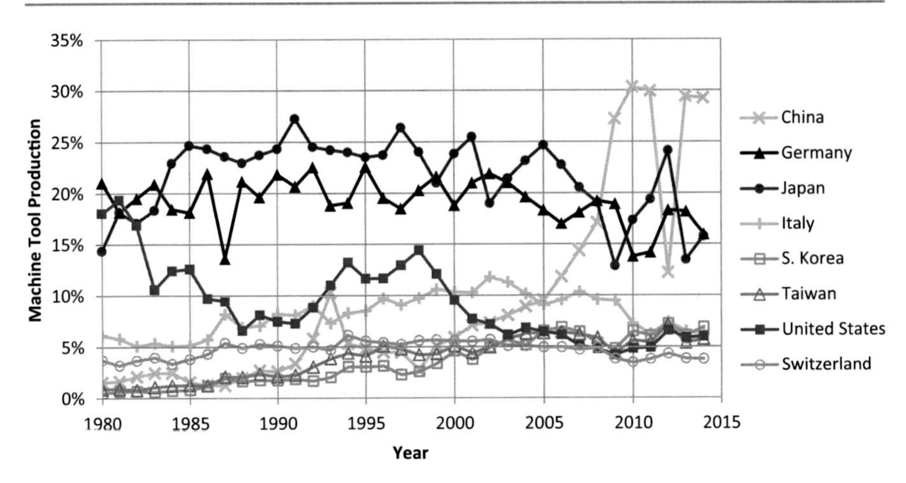

Fig. 1.2 Machine tool production by country from 1980 to 2014. (Source: Gardner)

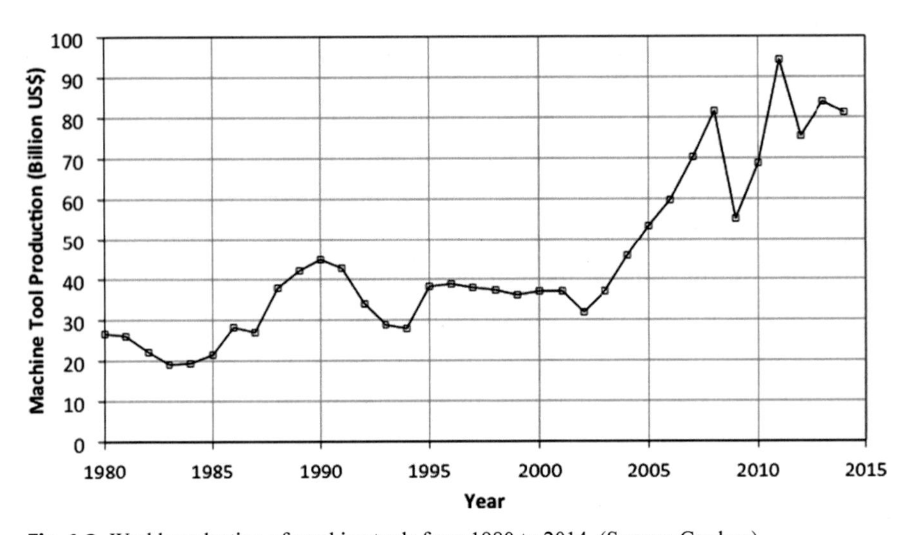

Fig. 1.3 World production of machine tools from 1980 to 2014. (Source: Gardner)

1.4 General Requirements in the Construction of Machine Tools

Machine tools are expected to fulfill the highest demands put upon them to implement technological advances in production. Apart from the purely functional capabilities that must be fulfilled, the ease of operation of the machine must be provided for its economic operation (attention to controls and layout). The adherence to a variety of legal requirements is a further constraint. The following problem areas arise out of these demands:

- Accuracy (geometric and kinematic) when statically, dynamically, and thermally loaded
- Retention of stability
- Automation
- Reliability
- Environnemental influences (energy use, carbon emission, noise, dust, corrosion, etc.)

Machine distortions affecting accuracy of performance during machining are influenced in reality by the relationship between static, dynamic, and thermal control factors, applicable to the particular type of machine. The accuracy of the product, the quality of surface finish, as well as the capacity and consequently the productivity of the machine, depend on these characteristics.

Further, the economic utilization of the machine is also determined through the degree of automation. Automation is not only dependent upon the cycle of the actual processes involved but also on the loading, unloading, and waste disposal requirements. The legal requirements made by the authorities to ensure the safety, energy, sustainability, and environmental acceptability of the machines play ever increasing roles. The aim is to reduce industrial accidents and to make the factory a safe and more pleasant place to work. However, it must not be overlooked that the efforts in that direction have, under certain circumstances, a detrimental effect on the productivity and economic efficiency of the machine. Again, a compromise is often necessary.

Considering the above constraints, the machine tool can only fulfill its function as a production device effectively if it is designed to suit the production demands that are expected of it as shown in Fig. 1.4. Firstly, the production process is determined, and according to the production demands and the appropriate processes, the fundamental parameters, such as the number and position of axes of movement and the functional capacities of the machining units, are determined. The expected accuracy of the work and the surface finish with which it is to be produced are determining factors for stiffness and rigidity of the machine components. The power provision will be dependent upon the expected workloads. The variety of the work to be performed, as well as the expected batch quantities, will govern the degree and type of automation provided.

Homework

1. To explore the importance of machining and machine tools in the society, you are asked to find the most up-to-date statistics of the following: (a) the number of employees in various sectors of the US manufacturing industry, (b) the momentary value of machine tool production, imports, and exports in (at least 5) different countries, (c) the momentary value of cutting machine tools and that of forming machine tools in various forms in a country other than the USA, (d) the evolution of machining speed over the years, (e) the evolution of machining accuracy over the years. Reference your citations clearly.

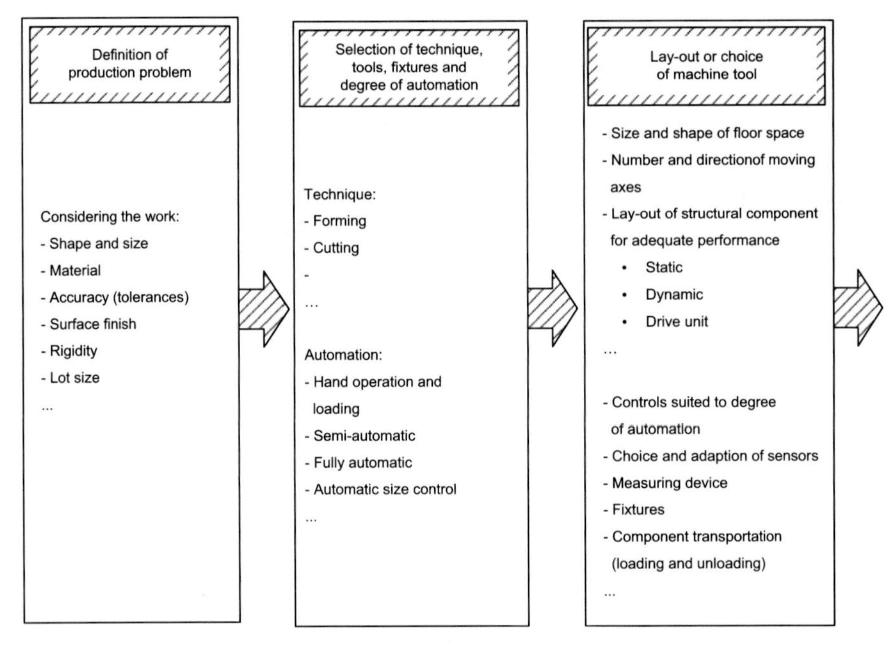

Fig. 1.4 Selecting or constructing machine tools for specific production operations

2. Summarize in 500–1000 words what you have learned from the world machine tools statistics above.
3. Identify and peruse an article in either mechanical engineering (ISSN 0025-6501) or manufacturing engineering (ISSN 0361-0853) on the topic of machining or machine tool industry as a whole in the USA or in the world, but not on any specific device or equipment. Summarize your lessons learned from and comments about the article in 500–1000 words.

2 Single-Point Cutting Processes

2.1 Motions in Machining

The principle used in all machining processes is to create the surface required by providing suitable relative motions between the cutting tool and the workpiece. The cutting edge or edges on the cutting tool remove a layer of work material. The removed material is often called a shaving or a chip. The simplest surfaces to generate are flat surfaces and internal or external cylindrical surfaces. For example, if a cutter is reciprocated back and forth along a straight line and a workpiece is incrementally fed beneath the tool in a direction perpendicular to the motion of the tool, a flat surface will be created on the workpiece. Similarly, a cylindrical surface can be generated by spinning the workpiece and feeding the tool parallel to the axis of workpiece. Therefore, in general, two kinds of relative motion must be provided by a machine tool. These motions are called primary motion and feed motion and are defined as follows.

The primary motion is the main motion provided by a machine tool to cause relative motion between the tool and workpiece. Usually, the primary motion absorbs most of the total power required to perform a machining operation. For example, in the case of flat surface creation, the motion of the tool relative to a stationary workpiece is the primary motion; in the case of cylindrical surface creation, the spinning motion of the workpiece relative to a stationary tool is the primary motion.

The feed motion, or the secondary motion, is a motion that may be provided to the tool or workpiece by a machine tool which, when added to the primary motion, leads to a repeated or continuous chip removal. It usually absorbs a small proportion of the total power required to perform a machining operation. In the case of flat surface creation, the motion of the workpiece relative to the ground is the feed motion; in the case of cylindrical surface creation, the linear motion of the tool relative to the machine frame is the feed motion.

Machine tools now are often controlled with complicated electronics allowing them to respond to more commands and making it possible to coordinate complex

S. Y. Liang, A. J. Shih, *Analysis of Machining and Machine Tools*,
DOI 10.1007/978-1-4899-7645-1_2

sets of motion. The functional definitions between turning, milling, drilling, and grinding are becoming blurred and each industry has its proprietary jargon for a given combination of functions or machine configuration. Commonly used metal-cutting machine tools, however, can still be divided into three groups depending upon the basic type of cutters used. These cutters can be (1) single-point tools, (2) multipoint tools, or (3) abrasive grits.

2.2 Machine Tools Using Single-Point Cutters

Single-point tools are cutting tools having one cutting part (or chip producing element) and one shank. They are commonly used in lathes, planers, shapers, boring mills, and similar machine tools. A typical single-point tool as used in cutting is illustrated in Fig. 2.1. One of the important tool angles when considering the geometry of a particular machining operation is the angle κ_r, the major cutting-edge angle. The thickness of the layer of material being removed is called the undeformed chip thickness, a_c, measured normal to the major cutting edge and the direction of chip flow. The feed, or feed engagement, f for all machine tools is defined as the displacement of the tool relative to the workpiece, in the direction of feed motion, per revolution of the workpiece or tool. Therefore,

$$a_c = f \sin \kappa_r, \tag{2.1}$$

which suggests that a_c increases with κ_r. Being just a bit ahead of the topic, it is understood that a larger undeformed chip thickness lowers the specific cutting energy (p_s, a term defined in Eq. (2.6)) thereby requiring less force/power to cut, which is

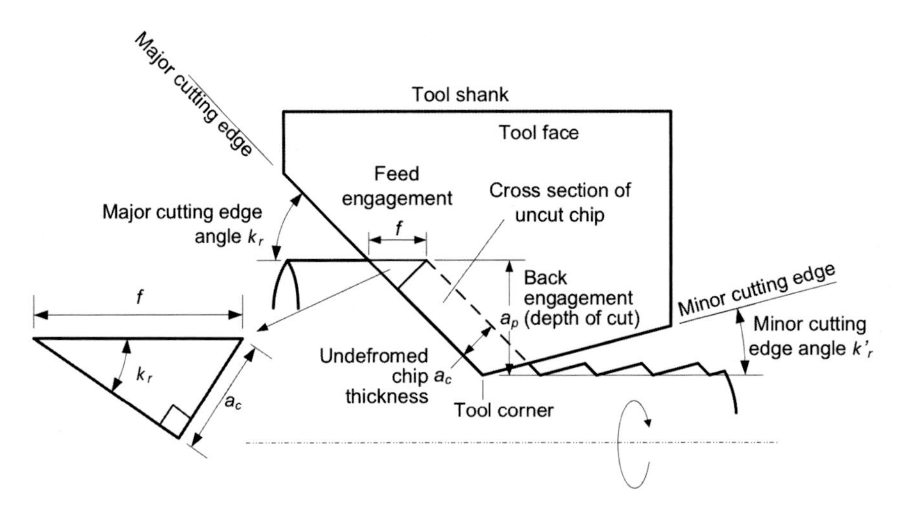

Fig. 2.1 Single-point tool operation

generally a merit. On the flip side, the surface finish is expected to become worse with greater κ_r, as indicated by Eq. (2.9) later.

For a small tool corner radius, the cross-sectional area A_c of the layer of material being removed can be approximated by

$$A_c = fa_p, \tag{2.2}$$

where a_p is back engagement (depth of cut).

Note that a portion of the workpiece is removed by the minor cutting edge near the corner. The minor cutting angle κ_r' usually is nonzero to avoid rubbing contact between the tool and the machined surface, but it assumes a small value that minimizes the height of machined surface profile as will be explained in the following section.

2.2.1 Lathe and Turning Process

In turning, a single-point cutting tool removes material from the external or internal surface of a workpiece rotating about its longitudinal axis. A typical lathe is shown in Fig. 2.2. It consists of a horizontal bed supporting the headstock, the tailstock,

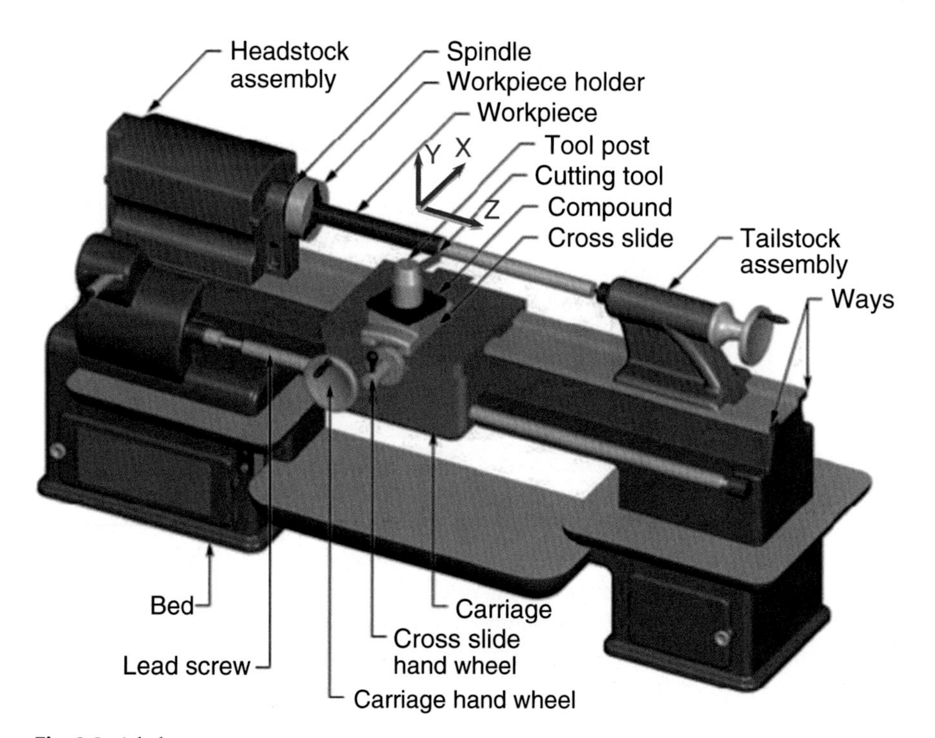

Fig. 2.2 A lathe

and the carriage. The workpiece is gripped at one end by a chuck mounted on the end of the main spindle of the machine and is supported at the other end by a center mounted in the tailstock. The tailstock can be clamped at various positions along the bed to accommodate workpiece of different lengths. Short workpiece needs only to be gripped by the chuck on one end because of the higher radial stiffness. With very long workpiece, a steady rest between the head and the tailstock is often used to minimize deflection. Next in complexity is the turret lathe. The turret has four or more sides, each with a tool. During turning, the operator rotates the turret so that more operations can be done in a single setup. The turret may be at the side, for outside diameter (OD) work, or at the end of the workpiece, for inside diameter (ID) work, replacing the tailstock.

Primary motion, the rotation of the workpiece, is provided by the movement of a series of gears driving the main spindle. The main spindle and the gears are all accommodated in the headstock. The single-point cutting tool is held in a tool post mounted on a cross slide, which in turn is mounted on the carriage. The carriage is driven along the bed ways in the Z direction by a leadscrew connected to the main spindle through a train of gears. The carriage can also remain stationary on the bed, while the gear train can be used to drive the tool holder across the carriage in the X direction using a leadscrew in lathe cross slide. The XYZ coordinate system shown in Fig. 2.2 is a convention that is commonly used in the computer numerical control (CNC) lathe, with the Z in the direction of the spindle rotation axis and −X in the direction toward the operator.

The rated size of a lathe indicates the largest diameter and length of workpiece it can handle. Thus a 14 in. lathe will swing a piece 14 in. in diameter over the bed ways. The swing is often designated by four digits, such as 1610. The first two designate the largest diameter that can be swung over the bed ways, and the last two designate the diameter over the cross slide.

The power feed drive on a lathe is always taken off the main spindle. In that way, the feed is always related to the speed of the spindle, and once chosen the feed setting remains constant regardless of the spindle speed. To turn a cylindrical surface of length l_w, the number of revolution of the workpiece is l_w/f, and the machining time t is given by:

$$t = \frac{l_w}{fn_w}, \tag{2.3}$$

where n_w is the rotational frequency of the workpiece in rpm.

Figure 2.3 illustrates five typical lathe operations. In Fig. 2.3a, which shows the geometry of a cylindrical-turning operation, the cutting speed at the tool corner is given by $\pi d_m n_w$, while the maximum value of cutting speed is given by $\pi d_w n_w$, where d_m is the diameter of the machined surface and d_w is the workpiece diameter. The maximum value of cutting speed is referred to as the cutting speed by convention. The average, or mean, cutting speed v is

$$v = \frac{\pi n_w(d_w + d_m)}{2} = \pi n_w(d_m + a_p). \tag{2.4}$$

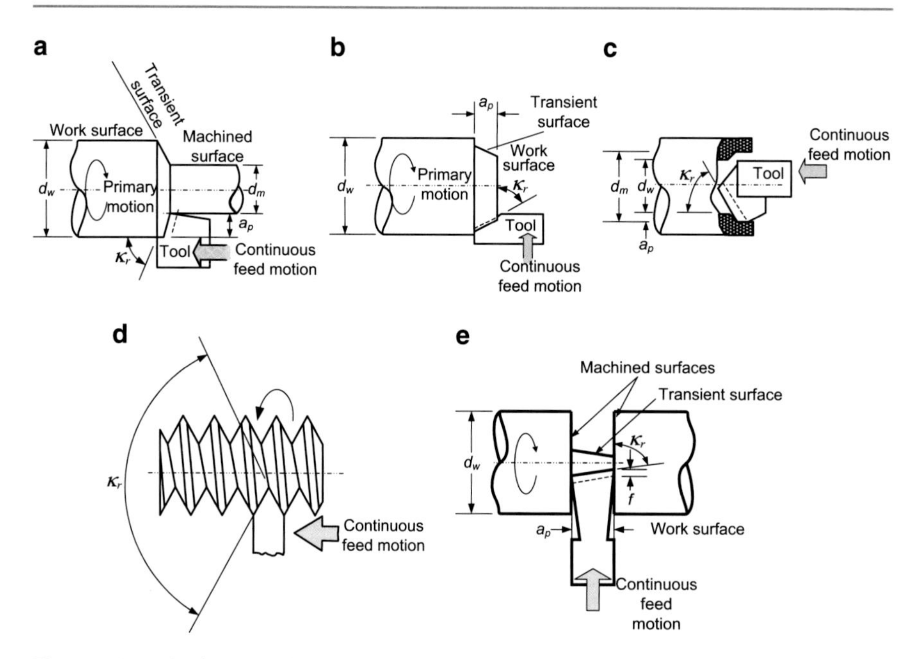

Fig. 2.3 Some basic types of lathe operations: **a** turning, **b** facing, **c** boring, **d** external threading, and **e** cutoff

The material removal rate (MRR) Z_w can be estimated as:

$$Z_w = A_c v = (a_p f)[\pi n_w (d_m + a_p)]. \tag{2.5}$$

For a given work material machined under given conditions, the energy required to remove a unit volume of material, p_s, can be measured. The p_s is referred to as the specific cutting energy. It is a material-dependent property although chip thickness in the cutting often has a bearing on its value. The power P_m required of the primary motion motor to perform any machining operation can be obtained from:

$$P_m = p_s Z_w. \tag{2.6}$$

Approximate values of p_s for various work materials and various values of the undeformed chip thickness are presented in Fig. 2.4.

Figures 2.5a and b show the surface finish in turning with a sharp nose cutter and with a finite-radius nose cutter, respectively. Part surface roughness is one of the common specifications requested of a machining process. Two of the widely used measures of surface roughness are the arithmetic average or centerline average roughness R_a:

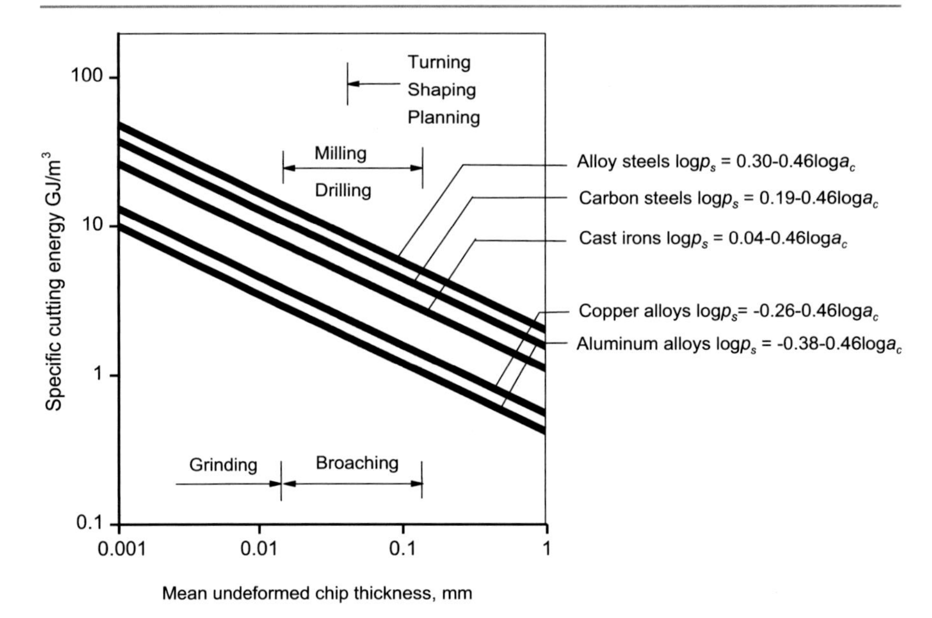

Fig. 2.4 Approximate values of the specific cutting energy (p_s) for mean undeformed chip thickness (a_c) with various materials and operation

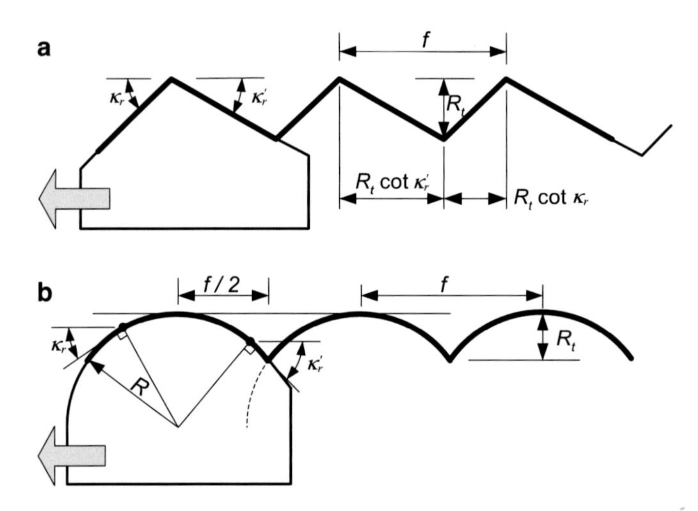

Fig. 2.5 Geometric surface profile in turning with **a** sharp nose edge and **b** finite nose radius

$$R_a = \frac{1}{L}\int_0^L \left| y(x) - \left(\frac{1}{L}\int_0^L y(x)dx \right) \right| dx \tag{2.7}$$

and the maximum (peak to valley) height roughness R_t:

$$R_t = \max_{0 \le x \le L}[y(x)] - \min_{0 \le x \le L}[y(x)], \tag{2.8}$$

with x being the coordinate along the surface, y the surface height, and L the distance of measurement. The surface profiles as shown in the figures are generated in a plane normal to the primary motion. For the geometry in Fig. 2.5a, the maximum (peak to valley) height roughness can be determined by

$$R_t = \frac{f}{\cot \kappa_r + \cot \kappa_r'}. \tag{2.9}$$

The R_a is referenced to the height center line; then according to the definition of Eq. (2.7), the portions below the center line are rectified, resulting in two triangles of height $R_t/2$ and base $f/2$. The two triangular areas add up to

$$R_a = \frac{1}{f}\left[2\left(\frac{1}{2}\right)\left(\frac{R_t}{2}\right)\left(\frac{f}{2}\right)\right] = \frac{R_t}{4} = \frac{f}{4\left(\cot\kappa_r + \cot\kappa_r'\right)}. \tag{2.10}$$

A radius R, is more common on the nose or corner of a cutting edge. This helps to provide a smoother finish as shown in Fig. 2.5b. Without further derivation, the R_t in this case is

$$R_t = (1 - \cos \kappa_r')R + f \sin \kappa_r' \cos \kappa_r' - \sqrt{2fR\sin^3 \kappa_r' - f^2 \sin^4 \kappa_r'}. \tag{2.11}$$

If only the circular arc generates the final profile the process can be described as finishing, as opposed to roughing. Since the minor cutting-edge angle is often much smaller than the major cutting-edge angle, the minor cutting edge is the one more likely to engage the workpiece with its straight section. Hence, the condition for finishing can usually be defined by the case when $\frac{f}{2} \le R \sin \kappa_r'$. For a finishing cut, the Eq. (2.11) reduces to

$$R_t \approx \frac{f^2}{8R}. \tag{2.12}$$

The general expression for R_a is somewhat complicated, but for finishing it is approximately

$$R_a \approx \frac{f^2}{32R}. \tag{2.13}$$

The roughness estimations discussed above are only theoretical lower limits because they ignore much material-related phenomena such as built-up edge, fracture, and vibrations that usually increase the roughness of a part. They serve their purpose as ways to plan a machining operation. For example, the optimal depth of cut or cutting speed can be determined based upon a specified surface finish or machining time, or the other way around, given the cutting process parameters the maximum achievable surface quality and production rate can be estimated from the above equations.

Example

A turning operation is carried out with the following conditions: Cutting tool is carbide with a major cutting-edge angle of 70°, a minor cutting-edge angle of 10°, and a tool nose radius of 3 mm. Workpiece is AISI 1045 steel bar 90 mm in diameter and 200 mm in length to be turned down to 75 mm diameter. Cutting parameters are spindle speed of 300 rev/min and feedrate of 1.5 mm/s. (a) Estimate the theoretical R_a and R_t roughness, in mm, for a new cutting edge and (b) estimate the power requirements to cut, in kW.

Solution

a. To determine R_a and R_t, we first need to check if it is a rough cut or finishing cut by checking if $\frac{f}{2} \leq R \sin \kappa_r'$.

First find $f = \text{feedrate} / \text{spindle speed} = 1.5\ (\text{mm/s}) / 5\ (\text{rev/s}) = 0.3\ \text{mm/rev}$; $2R \sin\kappa_r' = 2 \times 3 \times \sin(10°) = 1.042 \rightarrow$ Finishing cut, use Eqs. (2.12) and (2.13):

$$R_t = \frac{f^2}{8R} = (0.3)^2 / (8 \times 3) = 0.00375\ \text{mm},$$

$$R_a = \frac{f^2}{32R} = (0.3)^2 / (32 \times 3) = 0.000938\ \text{mm}.$$

b. To estimate the power requirements use Eq. (2.6) $P_m = p_s Z_w$

$$a_p = \text{depth of cut} = (d_w - d_m)/2 = (90 - 75)/2 = 7.5\ \text{mm}.$$

To obtain Z_w, use Eq. (2.5) $Z_w = \pi(d_w - a_p)n_w a_p f$

$$Z_w = \pi(d_w - a_p)n_w a_p f = \pi(90 - 7.5) \times 5 \times 7.5 \times 0.3 = 2916\ \text{mm}^3.$$

The chip thickness $a_c = f \sin \kappa_r = 0.3 \sin(70^\circ) = 0.282$ mm.

See p_s in Fig. 2.4 based on a_c: $\log p_s = 0.19 - 0.46 \log a_c \rightarrow p_s = 2.77$ GJ/m^3

$$P_m = Z_w p_s = 2916 \times 2.77 = 8076\ \text{W} = 8.08\ \text{kW}.$$

An operation in which a flat surface is generated by an engine lathe as shown in Fig. 2.3b is known as facing. The machining time t_m is given by

$$t_m = \frac{d_w}{2 f n_w}. \tag{2.14}$$

The maximum cutting speed v_{max} and the maximum metal removal rate $Z_{w,max}$, occurring at the beginning of the cut, are given by

$$v_{max} = \pi n_w d_w, \tag{2.15}$$

$$Z_{w,max} = \pi f a_p n_w d_w. \tag{2.16}$$

Example

With the use of a lathe, a facing operation is performed to shorten a 100 mm diameter aluminum bar by 5 mm. The lathe bit has a major cutting angle of 78° and a minor cutting angle of 5°. The spindle speed can be selected within 60–1200 rpm. It is required that the operation be completed within 30 s. We would like to accomplish the best part surface finish while meeting the throughput requirement. What is your recommendation on the minimum acceptable spindle motor power (in hp or in W) to accomplish this goal? Note that safety factors need not be considered.

Solution

Using Eq. (2.14) we can rearrange to find the feed: $t_m = \frac{d_w}{2 f n_w} \rightarrow f = \frac{d_w}{2 t_m n_w}$,

$n_w = 1200/60 = 20$ rps $\rightarrow f = 100/2(30)(20) = 0.083$ mm/rev. If the feed is greater than 0.083 mm, then the finish would be worse. If the feed was less than 0.083 mm then the machining time would exceed the time requirements. So the planned feed should be 0.083 mm.

Note that if $n_w < 20$ rps is used, f has to be greater than 0.083 mm to meet the time limit, then the finish worsened.

$$a_c = \text{chip thicness} = f \sin \kappa_r = 0.083 \sin(78^\circ) = 0.081 \text{ mm}.$$

Use Eq. (2.16), $Z_{w,max} = \pi f a_p n_w d_w = \pi \times 100 \times 20 \times 0.083 \times 5$ and $P_m = p_s Z_w$
from Fig. 2.4: $\log p_s = -0.38 - 0.46 \log a_c \rightarrow p_s = 1.32 \text{ J/mm}^3$

$$P_m = \pi \times 100 \times 20 \times 0.083 \times 5 \times 1.32 = 3442 \text{ W} = 4.62 \text{ hp}.$$

The process shown in Fig. 2.3c, termed boring, generates an internal cylindrical surface on a lathe. It is not to be used to create a new hole but is for the enlargement of an existing hole in the workpiece. Its MRR is

$$Z_w = \pi a_p f n_w (d_m + a_p). \tag{2.17}$$

The lathe operation illustrated in Fig. 2.3d is known as external threading or screw cutting. The primary motion generates a helix on the workpiece and is obtained by setting the gears that drive the leadscrew to give the required pitch of the machined threads. The last operation shown in Fig. 2.3e is referred to as parting or cutoff. It produces two machined surfaces simultaneously and is used when the finished workpiece is to be separated from the bar of material gripped in the chuck.

2.2.2 Boring Machine

This type of machine is needed mostly for heavy workpieces in which an internal cylindrical surface is to be machined. The principal feature of the machine is that the workpiece remains stationary during machining, and all the generating motions are applied to the tool as seen in Fig. 2.6.

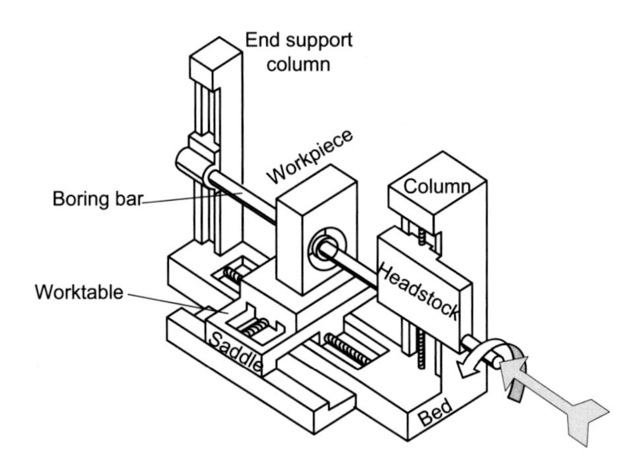

Fig. 2.6 Boring on a horizontal-boring machine

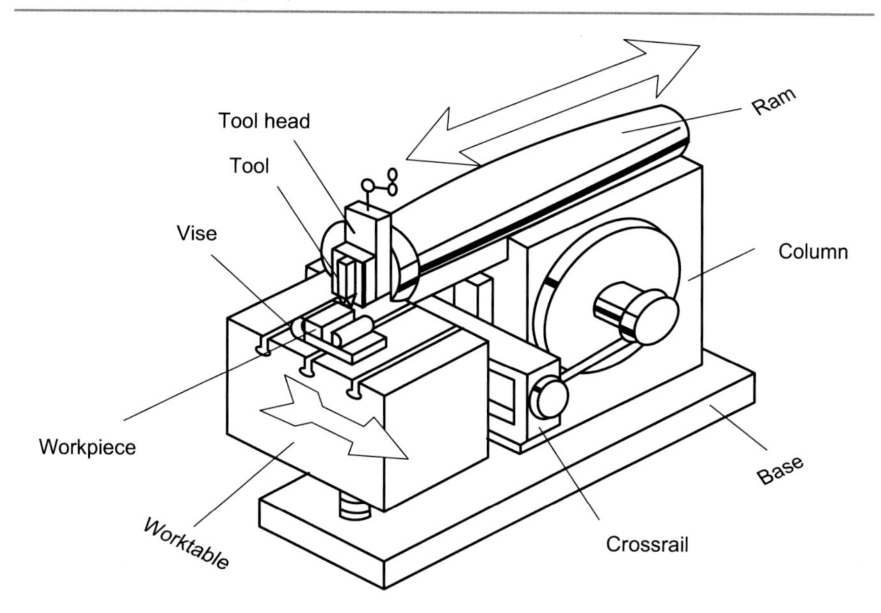

Fig. 2.7 Machining of a flat surface on a shaper

2.2.3 Shaping Machine

The primary motion on a shaping machine is linear as shown in Fig. 2.7. A single-point tool is gripped in a tool head mounted on the end of a ram which is made to move backward and forward. The cutting stroke is the forward stroke and the feed is applied to the workpiece in increments at the end of the return stroke of the ram by a ratchet-and-pawl mechanism driving the leadscrew in the cross rail.

The geometries when shaping horizontal, vertical, and inclined flat surfaces are shown in Fig. 2.8. For a surface of width b_w the machining time t_m is given by

$$t_m = \frac{b_w}{fn_r}, \tag{2.18}$$

where n_r is the frequency of cutting strokes (number of full back-and-forth round of strokes per minute) and f is the feed. The MRR Z_w will be given by

$$Z_w = fa_p v, \tag{2.19}$$

where v is the cutting speed and a_p is the back engagement as shown in Fig. 2.8. The undeformed chip thickness a_c will be given by

$$a_c = f \sin \kappa_r. \tag{2.20}$$

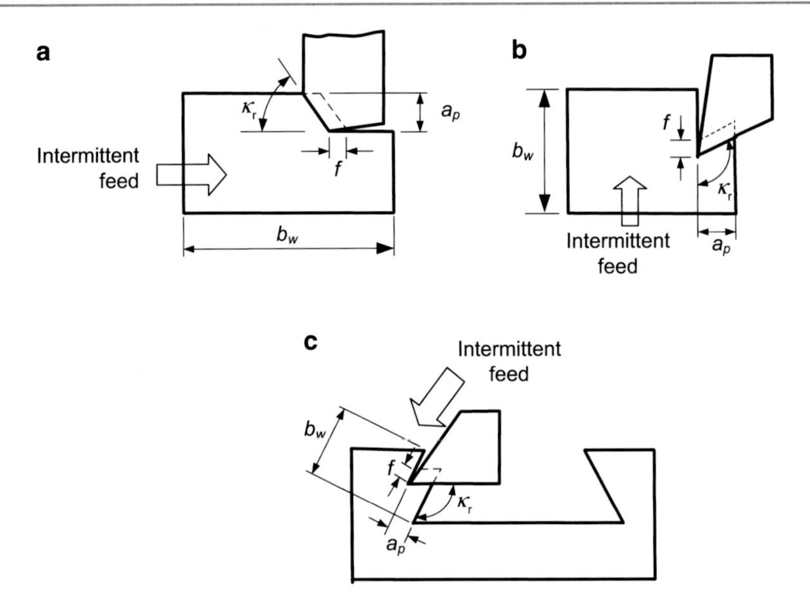

Fig. 2.8 Shaping for **a** horizontal, **b** vertical, and **c** inclined surfaces

> **Example**
> A horizontal shaping operation is planned to produce a flat and smooth (R_a=0.5 μm) surface on a 2024-T4 aluminum block of 80 mm by 80 mm by 80 mm. This will be done on a shaper that has 1.5 kW available at the forward stroke and with a high-speed steel (HSS) cutting edge with a primary cutting-edge angle of 80° and a minor cutting-edge angle of 5°. This cut will be made with a constant tool speed and feed to reduce the part height to 78 mm. (a) Choose a tool speed and feed that will meet the surface finish specification and make the cut as fast as possible without stalling the shaper motor and (b) estimate the time it will take to complete the cut you have planned.

Solution

a. Use Eq. (2.9) for R_a to get feed:

$$R_a = \frac{f}{4(\cot \kappa_r + \cot \kappa_r')} \rightarrow f = (0.0005)\,(4)\,(\cot 80^\circ + \cot 5^\circ) = 0.0232 \text{ mm.}$$

From Eqs. (2.19) and (2.6), we have $Z_w = fa_p v$ and $P_m = p_s Z_w \rightarrow P_m = p_s f a_p v$

$$a_p = 80 - 78 = 2 \text{ mm.}$$

From Fig. 2.4: $\log p_s = -0.38 - 0.46 \log a_c$, where $a_c = f \sin \kappa_r = 0.0232 \sin 80^\circ$ $\rightarrow p_s = 2.37$ J/mm^2

$$1.5 \text{ kW} = 1500 \text{ W} = (0.0232)(2)\, v\, (2.37) \rightarrow v = 13640 \text{ mm/s}.$$

b. Using Eq. (2.18) for machining time t_m, we have $t_m = b_w / (fn_w)$ where n_w is the workpiece feed frequency (s^{-1}). The distance traveled for the forward stroke is 80 mm and the return stroke is also 80 mm. Therefore, the total distance traveled for each passing stroke is 160 mm. Divide tool speed by the total distance for each passing stroke to get the workpiece tool frequency $n_w = v /$ Distance per stroke $= 13640/160 = 85.3 \;\; \text{s}^{-1} = 85.3$ Hz.

$$t_m = b_w / (fn_w) = 80 / (0.0232 \times 85.3) = 40.5 \text{ s}.$$

2.2.4 Planing Machine

The shaping machine is unsuitable for generating flat surfaces on very large parts because of limitations on the stroke and overhang of the ram. This problem is solved in the planing machine by applying the linear primary motion to the workpiece and feeding the tool at right angles to the motion as shown in Fig. 2.9.

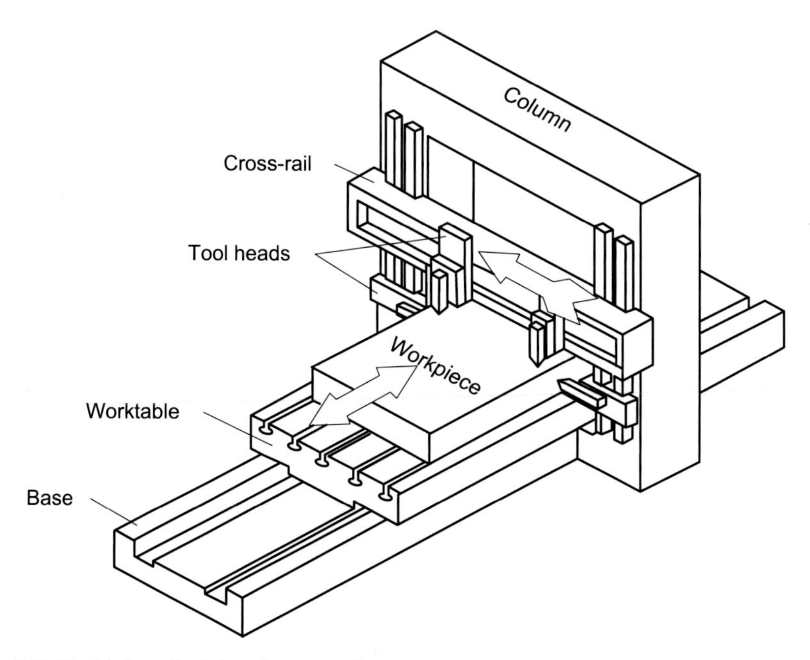

Fig. 2.9 Machining of a flat surface on a planer

Homework

1. A 2024-T4 aluminum cylinder of 64 mm diameter is turned on a lathe to reduce its diameter by 5 mm. The length to be machined on the cylinder is 105 mm. The lathe has 2 kW (2.68 hp; note that 1 hp = 745 W) available at the spindle and the HSS cutter has a major cutting-edge angle of 70° and a minor cutting-edge angle of 6°. Suppose the operation is to be done in one pass while a smooth surface with R_a less than 5 µm is desired, what should be the spindle speed and how much machining time would require?
2. The following figure shows the facing of a hollow cylinder with inner diameter d_i of 20 mm and outer diameter d_o of 50 mm. This operation is performed on a computer-controlled machine that reads the instantaneous tool position. Based on the tool position, the tool infeed per revolution is varied during the cut to maintain a constant MRR of 549 mm³/s throughout the process. As a_p of 1.5 mm is used, what is the total machining time (in s)? If a constant infeed process is exercised with a maximum MRR of 549 mm³/s at the beginning of cut, what is the total machining time required?

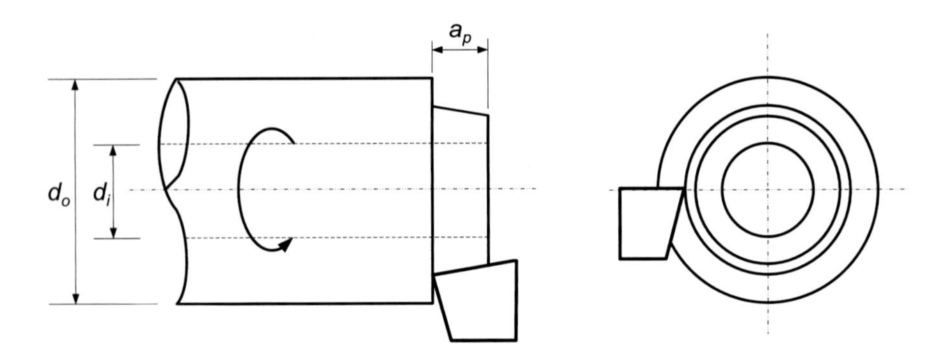

3. An inclined shaping is performed on cast iron workpiece as shown in the following figure. The cutting stroke is the forward stroke at a cutting velocity pointing out of paper. The intermittent feed, f, is applied at the end of the return stroke. What would be the power required if the cutting velocity is 65 mm/s, feed is 0.5 mm, depth of cut (a_p) is 10 mm, the major cutting-edge angle κ_r is 120°, and minor cutting-edge angle κ_r' is 5°? What would be the surface finish in R_t?

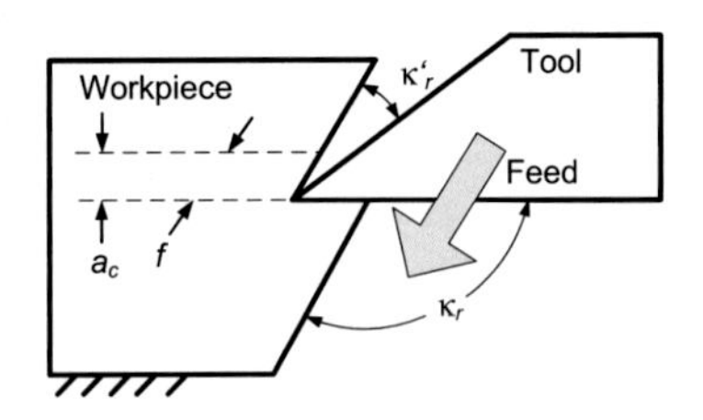

4. A 25 mm diameter cylindrical workpiece of carbon steel is turned to 23 mm with a HSS cutter of 75° major cutting-edge angle and 5° minor cutting edge angle. To achieve 300 mm length of cut within 2 min while maintaining a maximum-height-roughness of 10 μm or better, what is the lowest spindle speed (in rps) required? At that spindle speed what is the MRR (in mm^3/s)? You may assume the cutting edge is perfectly sharp with no tool nose radius.
5. An HSS turning bit with a major cutting-edge angle of 65° and a minor cutting edge angle of 10° is used on a 1 kW lathe to turn a 50 mm diameter cast iron bar. The maximum surface cutting speed that the tool can tolerate without overheating is 600 m/min. If a 100 mm long pass is to be completed within 1 min, what is the minimum R_a (in mm) achievable, and what is the largest (radial) depth of cut that can be executed? Do not use safety factors.

Multiple Point Cutting Processes 3

A multipoint tool can be regarded as a series of two or more cutting parts secured to a common body. The majority of multipoint tools are intended to be rotated and the workpiece is moved in a plane normal (as in milling) or parallel (as in drilling) to the axis of spindle rotation. Although the tool geometry and the resulting part features in multipoint machining are different from those in turning processes, the fundamental aspects of chip formation and cutting energy are very similar.

3.1 Milling Machine and Milling Process

Milling machines are some of the most versatile and productive machine tools in terms of their material removal rate (MRR). They are used most often to produce contour and planar surfaces. The process involves a prismatic workpiece or workpiece with sides. A rotating cutter with multiple cutting edges removes material. Either the cutter or workpiece can move or feed. If the spindle axis of rotation is normal to the machined surface, the process is called face milling as shown in Fig. 3.1. In face milling, the tool has cutting elements on both the tools face and periphery. If the spindle rotation is parallel to the machined surface, the process is called end milling or peripheral milling as shown in Fig. 3.2. In end milling, the cutting elements are on its periphery only. In all of these cases, the feed rate is in a plane normal to the spindle rotation.

The knee-and-column configuration is the simplest milling machine design. The workpiece is fixed to a bed on the knee, and the tool spindle is mounted on a column. Either or both can move. For very large workpieces, gantry or bridge-type milling machines are used. The two-column design gives greater stability to the cutting spindle(s).

Milling machines are generally classified as horizontal or vertical, depending on which way the spindle is mounted; which type to use hinges on; and what must be done to the workpiece? On vertical machines, the workpiece is secured to a horizontal table and therefore it is more stable than on horizontal machines. Cost

S. Y. Liang, A. J. Shih, *Analysis of Machining and Machine Tools*,
DOI 10.1007/978-1-4899-7645-1_3

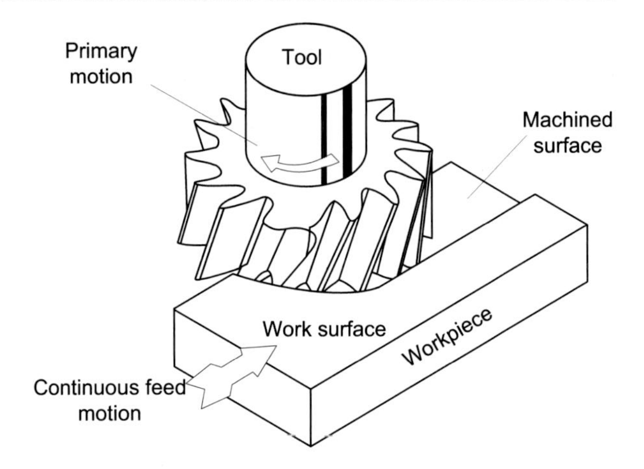

Fig. 3.1 Face milling

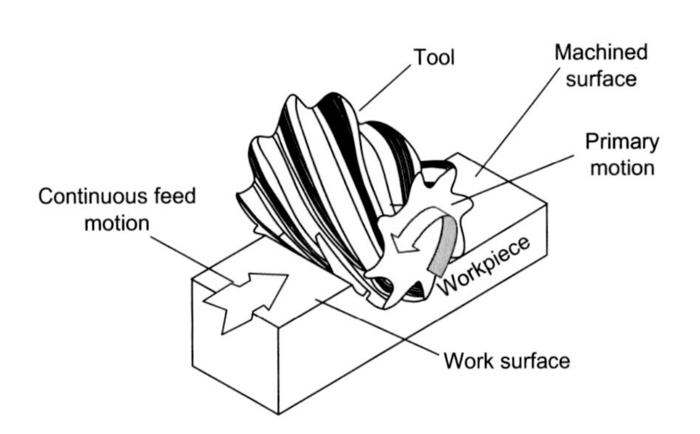

Fig. 3.2 End milling

of operation may be lower because less power is required for axis motion on the table plane. Mounting of the larger workpiece is easier and it is easier to achieve accuracy because of the rigidly mounted workpiece. It is also better for the plate-type workpiece.

In a horizontal machine, cutting costs may go higher because the workpiece is fastened to a vertical fixture. However, chip removal is easier and it is usually easier to work on larger complex parts. There is also less restriction on vertical height of the workpiece.

There are also "universal" machines with heads that can be rotated so the machining is done either horizontally or vertically. This provides machine versatility, but adds cost and another source of head-positioning error.

The geometry for face milling and angles for a face milling cutter are shown in Fig. 3.3. A cross section of the chip at its thickest point, where $\theta = 0°$ and the cutting edge parallel to the feed rate direction, is shown in Fig. 3.4. The cross-hatched chip width b is

$$b = \frac{d}{\cos\kappa}, \tag{3.1}$$

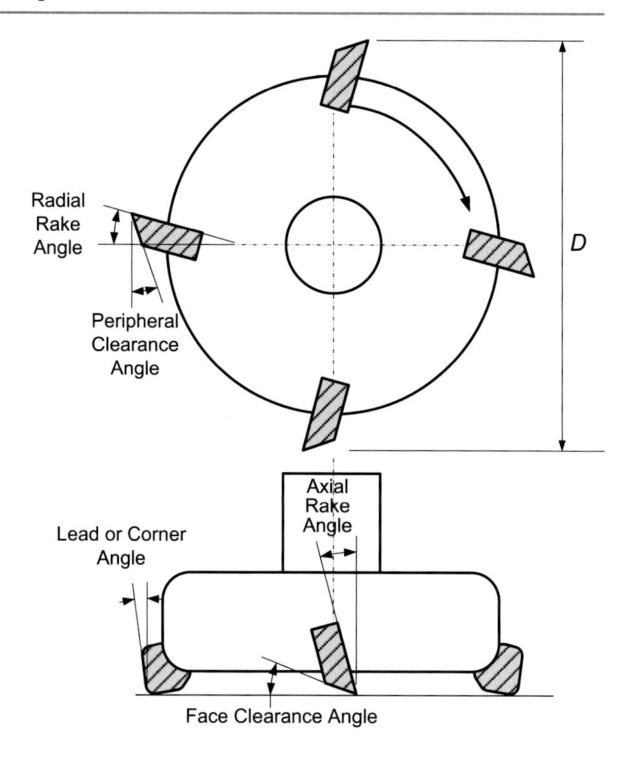

Fig. 3.3 Geometry of a face mill

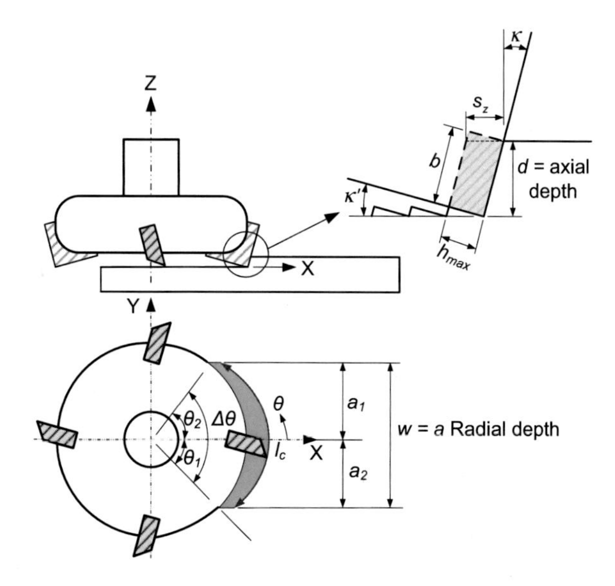

Fig. 3.4 Chip thickness, contact length, and surface roughness in face milling

where d is the axial depth of cut and κ is the lead or corner angle.

From the top view of the cutter, it is seen that the chip thickness, measured radially, varies throughout the cutter's rotation. The maximum chip thickness in face milling can be found to be

$$h_{max} = s_z \cos\kappa, \tag{3.2}$$

where s_z is the feed per tooth. If the lead angle is zero, $h_{max} = s_z$. Thus, the instantaneous chip thickness in face milling is

$$h(\theta) = h_{max} \cos\theta = s_z \cos\kappa \cos\theta, \tag{3.3}$$

where θ is the angular coordinate system with $\theta = 0$ being in the feed rate direction, θ_1 being the entry angle, and θ_2 being the exit angle. Note that both θ_1 and θ_2 are defined to be positive by convention. The length of the discontinuous chip in face milling depends on the cut geometry as follows:

$$l_c = \frac{D\Delta\theta}{2} = \frac{D}{2}\left[\sin^{-1}\left(\frac{2a_1}{D}\right) + \sin^{-1}\left(\frac{2a_2}{D}\right)\right], \tag{3.4}$$

where, as illustrated in Fig. 3.4, D is the diameter of the face mill, $\Delta\theta$ is the angle span of contact between the face mill and workpiece, and a_1 and a_2 are the radial depth in the entry and exit sides of the workpiece, respectively. From this, the average chip thickness can be defined based on

$$\begin{aligned} h_{ave} &= \frac{\int_{-\theta_2}^{\theta_1} h(\theta)\frac{D}{2}\,d\theta}{l_c} = \frac{Dh_{max}}{2l_c}\int_{-\theta_2}^{\theta_1}\cos\theta\,d\theta = \frac{Dh_{max}}{2l_c}(\sin\theta_2 + \sin\theta_1) \\ &= \frac{s_z \cos\kappa[2(a_1 + a_2)]}{D\left[\sin^{-1}\left(\frac{2a_2}{D}\right) + \sin^{-1}\left(\frac{2a_1}{D}\right)\right]}. \end{aligned} \tag{3.5}$$

Being able to estimate the average chip thickness is of particular importance when using the specific cutting energy method to estimate power and forces.

The MRR in face milling (or milling as a whole) is not constant as it is in single point turning processes. The time variation of MRR is a result of intermittent engagement and the overlapping between teeth. Accordingly, the machining power also varies with time. To calculate the total power required on an estimation basis, an average MRR is often used in association with a magnification factor, that is, $P_m = p_s Z_{w,ave} n$, where $Z_{w,ave} = a d s_z N N_t$ and the magnification factor n can range from 1.5 to 2. The variable a is the radial depth of contact, N is the tool rotational speed, and N_t is the number of teeth. The specific cutting energy p_s is estimated based on the average chip thickness given in Eq. (3.5).

In face milling, the theoretical surface profile, measured parallel to the feed rate and coincident with the cutter axis, is equivalent to the profile parallel to the axis of rotation in turning with $f = s_z$, $\kappa_r = (\pi/2) - \kappa$, and $\kappa_r' = \kappa'$. Therefore, when the cutter corner radius, $R=0$, the surface roughness R_t and R_a in face milling are:

$$R_t = \frac{s_z}{\tan\kappa + \cot\kappa'} \tag{3.6}$$

$$R_a = \frac{s_z}{4(\tan\kappa + \cot\kappa')}. \tag{3.7}$$

In a similar way, if the nose radius on the cutting edges is not zero, the roughness R_t and R_a are:

$$\begin{aligned} R_t &= (1-\cos\kappa')R + s_z \sin\kappa' \cos\kappa' - \sqrt{2 s_z R \sin^3\kappa' - s_z^2 \sin^4\kappa'} \\ &\approx \frac{s_z^2}{8R}, \text{ for } \frac{s_z}{2} \le R\sin\kappa' \end{aligned} \tag{3.8}$$

$$R_a \approx \frac{s_z^2}{32R}, \text{ for } \frac{s_z}{2} \le R\sin\kappa'. \tag{3.9}$$

Again, these are theoretical roughness values based on kinematics. They will not be correct if the spindle is not true, that is, the spindle rotates about an axis different from its geometric axis.

Kinematics to estimate cutting time and tool life for face milling is more complicated than for turning because there is a transient portion at the beginning and end of a pass where an edge is not continuously cutting like it is in turning. Referring to Fig. 3.4, the time spent in producing chips from a workpiece of length L is

$$t = \frac{L + \frac{D}{2}\left(1 - \cos\left(\max[\theta_1, \theta_2]\right)\right)}{s_z N N_t}. \tag{3.10}$$

Example

A cast iron part is to be slotted with a 55 mm four-flute cutter as shown in the following figure. The cutter inset has a corner angle of 20°, a face cutting edge angle of 10°, and a corner radius of 1.5 mm. The milling machine can deliver a maximum power of 1600 W at the spindle. The requirement herein is to achieve a machined surface finish better than 0.2 μm in R_a. Estimate the minimum machining time required to slot the entire part at axial depth of cut of 1.5 mm. Note that a safety factor of 2.5 should be used with the specification of finish.

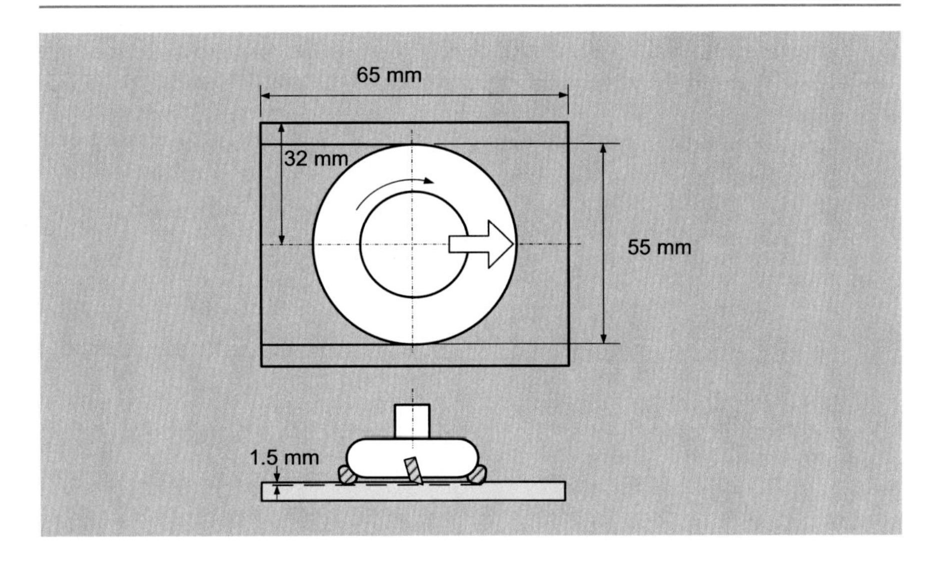

Solution

Equation (3.9): $s_z^2 \le 32RR_a = 32(1.5)\left(\dfrac{0.0002}{2.5}\right) = 0.061^2$

Use $s_z = 0.06$ mm (Check if $s_z < 2(1.5)\sin 10^\circ = 0.5$ mm. It is fine.)

Average $Z_{w,ave} = ads_z NN_t = (55)(1.5)(0.06)(4)\text{N} = 19.8\ \text{N}$

$$h_{ave} = \frac{s_z \cos\kappa\left[2\left(a_1 + a_2\right)\right]}{D\left[\sin^{-1}\left(\dfrac{2a_2}{D}\right) + \sin^{-1}\left(\dfrac{2a_1}{D}\right)\right]} = s_z \frac{2}{\pi}\cos\kappa$$

$$= 0.036 \text{ mm (since } \theta_1 = \theta_2 = 90^\text{o}\text{)}.$$

From Fig. 2.4, $p_s = 5.06\ \text{J/mm}^3$

$$\text{Max power} = (1600)/(\text{Safety factor} = 2.5) > p_s Z_w \times (\max Z_w \text{ correction} = 1.5) \rightarrow N \le 4.26 \text{ rps}$$

$$\text{Machining time } t = \frac{L + D/2}{s_z NN_t} = \frac{65 + 55/2}{0.06(4)(4.26)} = 90.5 \text{ s.}$$

Example

A block of cast iron workpiece, with a length of 5 in., is face milled by a 2 in. diameter four-flute carbide insert cutter as shown in the following figure. The cutter inserts all have a lead angle of 45° and a face cutting edge angle of 8°.

It is required that the entire 5 in. block should be machined within 20 s in one pass while the resulting finish on the machined surface shall not exceed 1×10^{-4} in. R_a. What are the upper and/or lower bounds for the acceptable spindle speed (in rps) in this case? At these speed limits, what are the average cutting powers? You may assume that the cutter inserts have perfectly sharp corners.

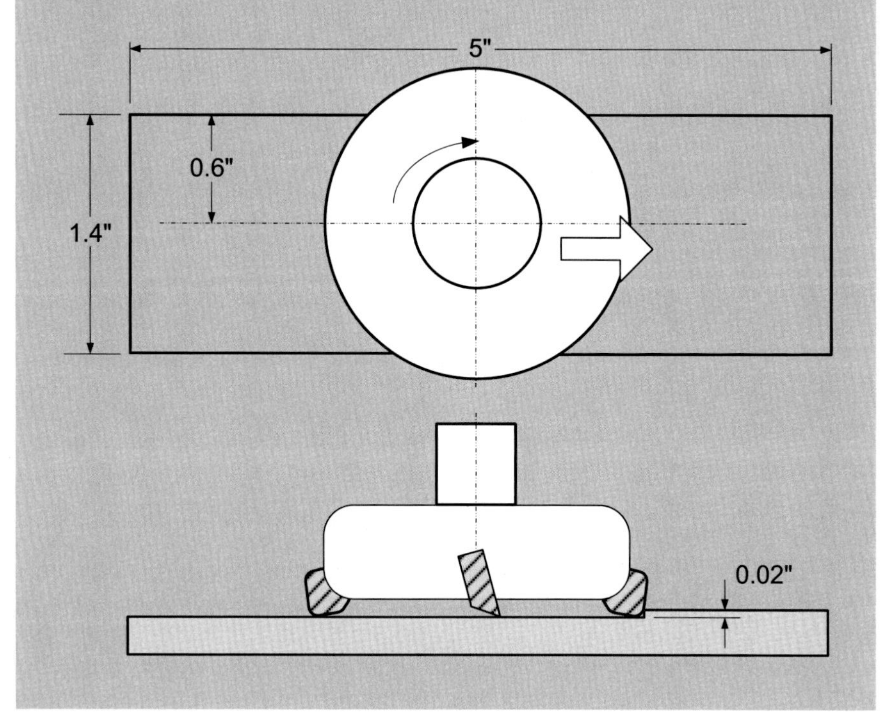

Solution

From Eq. (3.7): $s_z = 4R_a(\tan\kappa + \cot\kappa')$

$$\theta_1 = \sin^{-1}\left(\frac{0.6}{2/2}\right) = 37^{\circ}$$

$$\theta_2 = \sin^{-1}\left(\frac{1.4 - 0.6}{2/2}\right) = 53^{\circ}.$$

From Eq. (3.10): $t = \dfrac{L + \frac{D}{2}\left(1 - \cos\left[\max\left(\theta_1, \theta_2\right)\right]\right)}{\left(v_f = s_z N_t N\right)} \Rightarrow N = \dfrac{L + \frac{D}{2}(1 - \cos 53^\circ)}{R_a(4)(\tan\kappa + \cot\kappa')tN_t}$

$$= \frac{5 + \frac{2}{2}(0.4)}{(10^{-4})\,4\,(\tan 45^\circ + \cot 8^\circ)\,20\,(4)} = 20.8 \text{ rps},$$

which is the lower bound for the spindle speed. There is no upper bound involved in this case since there is no mention on the available motor power.

At this speed, $s_z - \left(10^{-4}\right)4\left(\tan 45^\circ + \cot 8^\circ\right) = 3.24 \times 10^{-3}$ in., and from Eq. (3.5):

$$h_{ave} = s_z \cos\kappa \frac{2\,(a_1 + a_2)}{D\,(\theta_1 + \theta_2)} = \frac{(3.24 \times 10^{-3})\cos 45^\circ\,(2)1.4}{2\,(37^\circ + 53^\circ)}$$

$$= 2.04 \times 10^{-3} \text{ in.} = 0.052 \text{ mm.}$$

From Fig. 2.4, the specific cutting power is about 4.27 GJ/m^3 (or J/mm^3)

$$P_m = p_s Z_w = (4.27 \text{ GJ/m}^3)[0.02(1.4)3.24 \times 10^{-3}(4)20.8(0.0254^3)] = 528 \text{ W.}$$

Figure 3.5 shows the geometry of an end mill. The primary cutting edges are along the flutes of the cutter, with the secondary cutting edges on the bottom. The helix angle, α, of the flutes helps to even out the force variation that the cutter sees, by having a cutting edge enter the workpiece gradually. The concavity angle, κ', which defines the secondary cutting edge, can have a major effect on the roughness

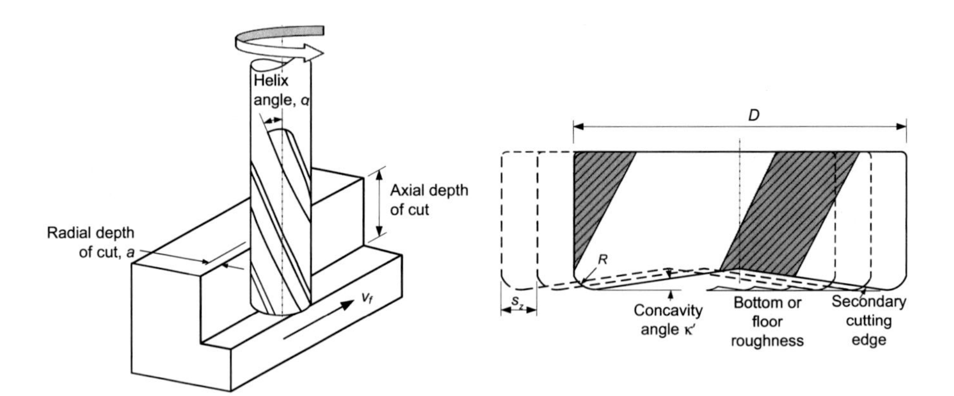

Fig. 3.5 Geometry of an end mill

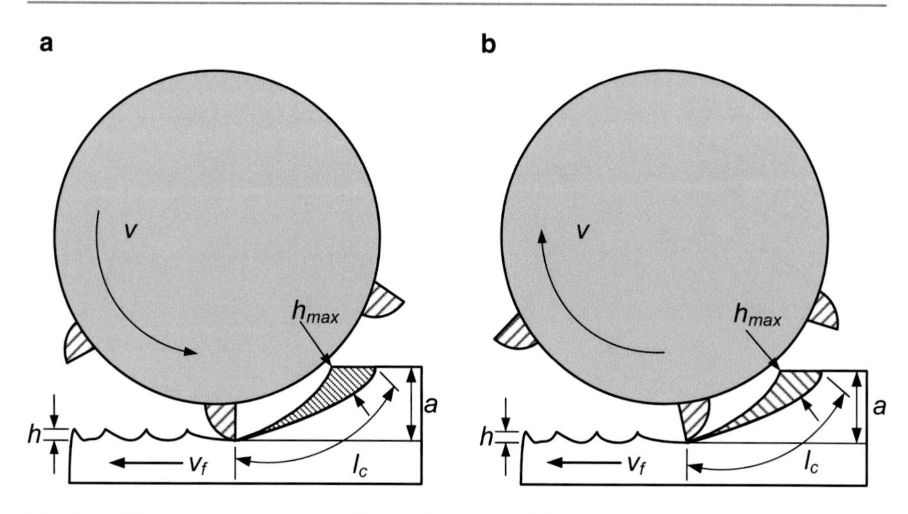

Fig. 3.6 Difference between **a** up milling and **b** down milling

of the floor surface. The wall surface, on the other hand, develops a roughness profile that is a function of the cutter diameter, the feed per tooth s_z, and feed rate v_f.

There are two possible end milling configuration depending on the tool velocity relative to the workpiece. In down or climb milling (Fig. 3.6b), the feed motion is given in the direction of cutter rotation while in conventional. In up milling (Fig. 3.6a), the workpiece is fed in a direction opposite to the cutter motion. Up milling usually provides a longer chip due to a larger range of tool–work engagement. For the same amount of MRR, a longer chip implies a smaller chip thickness, therefore, a smoother surface with less scalloping. However, in up milling the workpiece tends to be lifted due to the existence of an upward cutting force component; thus, the machine structure is more likely to be unstable. Cutting temperature is usually higher and a cutter usually lasts shorter in up milling.

Figure 3.7 represents a cross section along the cutter's path, and the chip contact length in end milling. It is seen that the chip length is

$$l_c = D\frac{\Delta\theta}{2}, \tag{3.11}$$

in which $\Delta\theta$ is the range of angular engagement determined from

$$\Delta\theta = \cos^{-1}\left(1 - \frac{a}{D/2}\right). \tag{3.12}$$

The chip thickness, measured radially, increases from zero to a maximum value in up milling, and it decreases from a maximum value to zero in down milling. For either configuration, the maximum chip thickness is

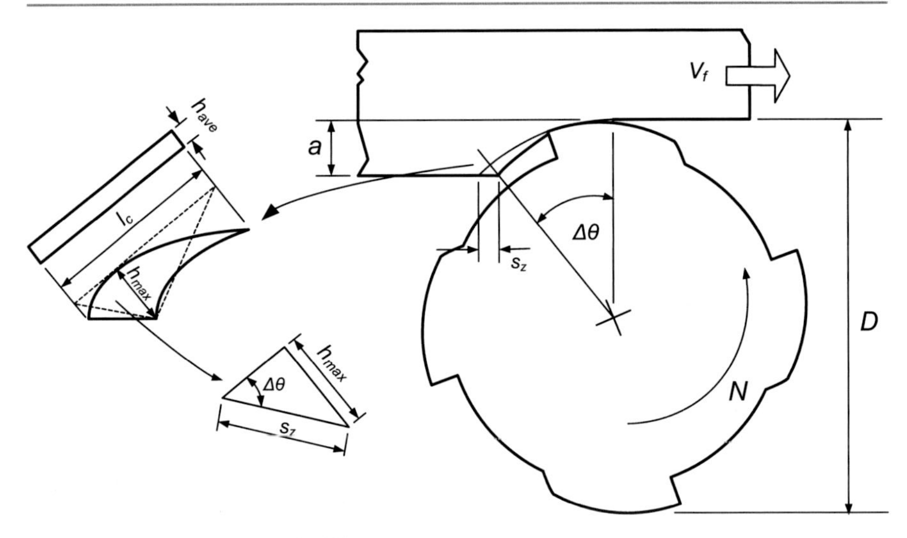

Fig. 3.7 Chip geometry in end milling

$$h_{\max} = s_z \sin \Delta\theta = s_z \sin\left[\cos^{-1}\left(1 - \frac{2a}{D}\right)\right]. \tag{3.13}$$

The average chip thickness is the thickness of a rectangular cross section of length l_c that has the same cross section as the triangular cross section, that is, $h_{ave}l_c = (h_{max}l_c)/2$, as seen in Fig. 3.7, so that

$$h_{ave} = \frac{h_{\max}}{2}. \tag{3.14}$$

This expression can be used to estimate the cutting force in end milling using specific cutting energy methods with given cutting parameters.

Example

A slab milling operation has the following specifications: Cutter: 100 mm diameter ten-flute HSS cutter. Workpiece: 1045 steel 150 mm wide. Cutting configuration: down milling with $N=100$ rev/min, $a=15$ mm, $v_f=4$ mm/s. Estimate the average required power, in W, to make this cut with a fresh cutter.

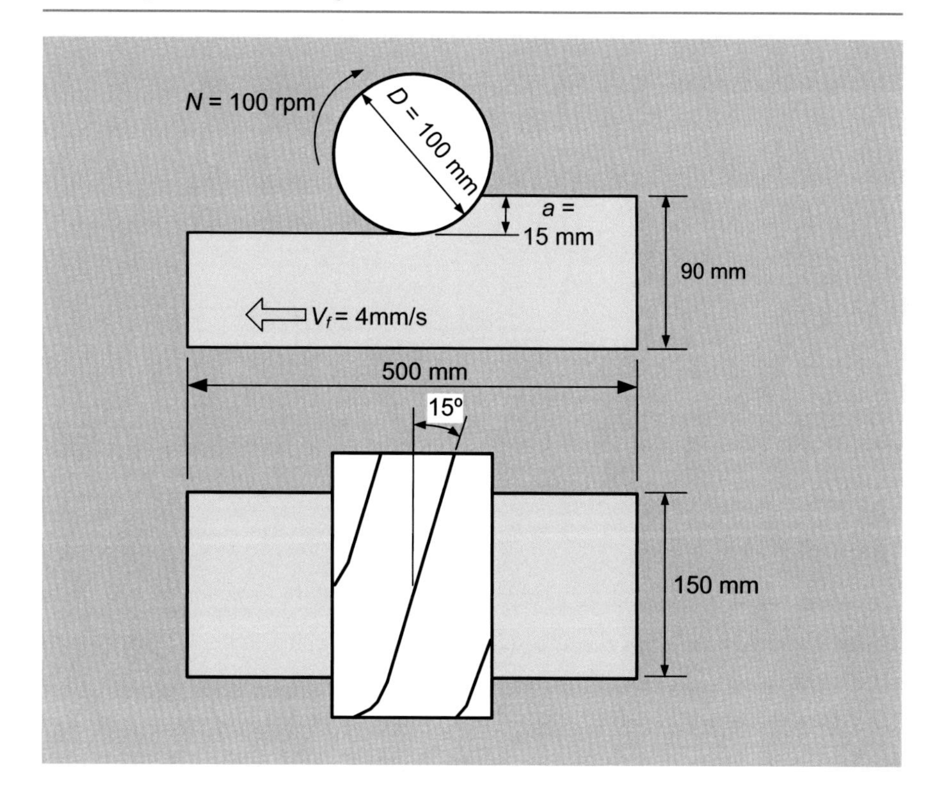

Solution

$$s_z = \frac{v_f}{NN_t} = \frac{4}{\frac{100}{60}(10)} = 0.24 \text{ mm}$$

$$\Delta\theta = \cos^{-1}\left(1 - \frac{a}{D/2}\right) = \cos^{-1}\left(1 - \frac{15}{100/2}\right) = 45.6^\circ$$

$$h_{ave} = \frac{1}{2} s_z \sin\Delta\theta = \frac{1}{2}(0.24)(\sin 45.6^\circ) = 0.086 \text{ mm.}$$

From Fig. 2.4, $p_s = 4.8 \text{ GJ/m}^3$ with an uncut chip thickness of 0.086 mm.

$$Z_w = awv_f = 15(150)4 = 9000 \text{ mm}^3/\text{s} = 9 \times 10^{-6} \text{ m}^3/\text{s}.$$

The average required power is $P_s = p_s Z_w = (4.8 \times 10^9)(9 \times 10^{-6}) = 43.2 \text{ kW}$.

The theoretical roughness for both the floor and wall in end milling can be estimated. Beginning at the floor, if the corner radius $R = 0$ in Fig. 3.5, the roughness of the floor depends only on the concavity angle κ' and the feed per tooth s_z, that is, $\kappa = 0$

$$R_t = \frac{s_z}{\cot \kappa'} \tag{3.15}$$

$$R_a = \frac{s_z}{4\cot \kappa'}. \tag{3.16}$$

If $R \neq 0$,

$$\begin{aligned} R_t &= (1-\cos\kappa')R + s_z \sin\kappa' \cos\kappa' - \sqrt{2s_z R \sin^3\kappa' - s_z^2 \sin^4\kappa'} \\ &\approx \frac{s_z^2}{8R}, \text{ for } \frac{s_z}{2} \leq R\sin\kappa'. \end{aligned} \tag{3.17}$$

$$R_a \approx \frac{s_z^2}{32R}, \text{ for } \frac{s_z}{2} \leq R\sin\kappa'. \tag{3.18}$$

From Fig. 3.6, profiles taken on the wall in end milling in the direction of feed can be determined by considering the effective nose radius of cutter to be $R = \frac{D}{2} \pm \frac{s_z N_t}{\pi}$, while N_t is the number of teeth on the cutter, with the plus used for up milling and the minus for down milling. Note that $\frac{s_z N_t}{\pi}$ represents the tool travel distance during half cycle of rotation. Thus,

$$R_t \approx \frac{s_z^2}{8\left(\frac{D}{2} \pm \frac{s_z N_t}{\pi}\right)} \tag{3.19}$$

$$R_a \approx \frac{s_z^2}{32\left(\frac{D}{2} \pm \frac{s_z N_t}{\pi}\right)}. \tag{3.20}$$

The total time spent producing chips is (see figure below)

$$t = \frac{\frac{D}{2}\sin\Delta\theta + L}{v_f}. \tag{3.21}$$

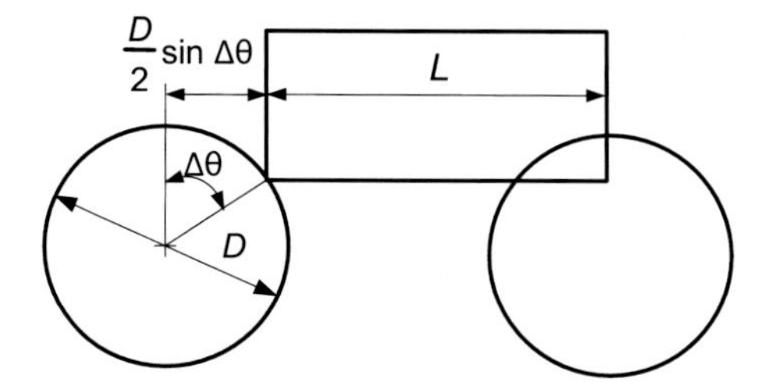

Example

In a conventional end milling process a four-flute, 20 mm diameter cutter is to machine a cast iron workpiece. The radial depth of cut is 0.5 mm and the axial depth of cut is 35 mm. The goal is to cut the 200 mm long rectangular workpiece in one pass in less than 30 s and to produce a surface roughness R_a less than 2 μm along the peripheral wall of the cutter. Determine the requirement on the spindle speed (in rpm). Is this requirement a minimum acceptable or a maximum acceptable spindle speed? If machining is carried out at this particular spindle speed, what is the average power (in W) needed to cut?

Solution

From Eq. (3.20):

$$R_a < \frac{s_z^2}{32\left(\frac{D}{2}+\frac{s_z N_t}{\pi}\right)} \Rightarrow 0.002 < \frac{s_z^2}{32\left(\frac{20}{2}+\frac{s_z(4)}{\pi}\right)} \Rightarrow s_z < 0.829 \text{ mm/tooth}$$

$$\Delta\theta = \cos^{-1}\left(1-\frac{a}{D/2}\right) = \cos^{-1}\left(1-\frac{0.5}{20/2}\right) = 18.2^\circ.$$

$$\text{From Eq. (3.21): } t < \frac{\frac{D}{2}\sin\Delta\theta + L}{s_z N_t N} \Rightarrow 0.5 < \frac{\left(\frac{20}{2}\right)\sin 18.2^\circ + 200}{(0.829)N(4)} \Rightarrow N > 122.5 \text{ rpm.}$$

The requirement is a minimum acceptable spindle speed. If a lower spindle speed is applied, the total machining time would be longer.

From Eqs. (3.13) and (3.14),

$$h_{ave} = \frac{1}{2}s_z \sin\left(\cos^{-1}\left(1-\frac{a}{D/2}\right)\right) = \frac{1}{2}(0.829)\sin\left(\cos^{-1}\left(1-\frac{0.5}{20/2}\right)\right) = 0.128 \text{ mm}$$

See Fig. 2.4 for cast iron workpiece at 0.128 mm average chip thickness, $p_s = 2.8$ J/mm^3.

$$\text{Average power} = p_s z_w = p_s a d s_z N_t N = (2.8)(0.5)(35)(0.829)(4)(122.5)$$
$$= 19{,}904 \text{ J/min} = 331 \text{ W}.$$

3.2 Drilling Process

Drilling machine produces internal cylindrical surface (holes) of moderate accuracy in terms of position, roundness, and straightness. It can be done on engine lathes or milling machines, in addition to dedicated drill presses or special stations on a transfer line. Holes made by drilling are often used for mechanical fasteners such as bolts or rivets. Drilling is a preliminary step for processes like tapping, boring, or reaming, and it is the machining process most frequently used on composite materials.

The most familiar way of drilling is done with a rotating tool that is fed into a stationary workpiece. However, the workpiece may rotate while the drill remains fixed. Of the drills used, the multiflute twist drill is the most common. The flutes on a twist drill are helical (formed by twisting, hence the name) and are not designed for cutting, as was the case for the helix on an end mill. The flute helix provides a way to remove chips from the hole while cutting occurs along the lips of the drill. The twist drill as shown in Fig. 3.8 has two cutting edges, each removing equal amount of the work material.

The cutting speed for drilling is defined as the rotational motion at the periphery of the drill and is given as:

$$v = \pi a_p N. \tag{3.22}$$

Note that N is the spindle rotational frequency and the depth of cut a_p in this case is the drill bit diameter. Feed rate v_f for drilling is the axial advancing speed into the workpiece

$$v_f = fN, \tag{3.23}$$

while f is the feed per revolution. The undeformed chip thickness a_c can be estimated as

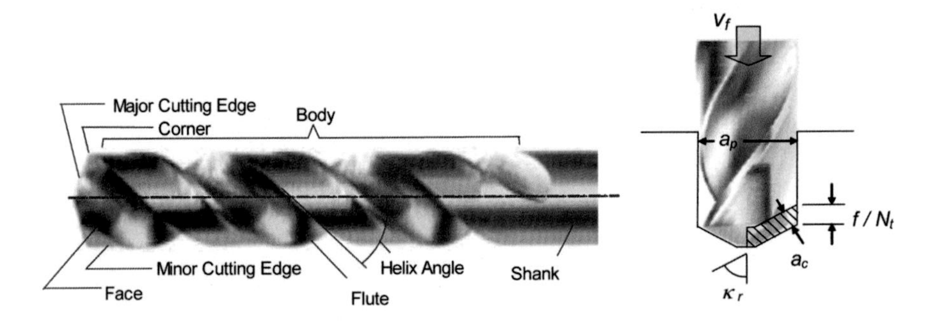

Fig. 3.8 Twist drill $a_c = a_f \sin \kappa_r$

$$a_c = \frac{f}{N_t}\sin\kappa_r, \tag{3.24}$$

where κ_r is termed the major cutting-edge angle in drilling, which is half of the drill point angle. The machining time to cut a hole of length l_w is

$$t = \frac{l_w + \frac{a_p}{2}\cot\kappa_r}{fN} = \frac{l_w + \frac{a_p}{2}\cot\kappa_r}{v_f}. \tag{3.25}$$

The MRR Z_w is

$$Z_w = \frac{\pi v_f a_p^2}{4}. \tag{3.26}$$

The drilling power is then

$$P = p_s Z_w = p_s \frac{\pi v_f a_p^2}{4} \tag{3.27}$$

from which the drilling torque can be estimated as

$$T = \frac{P}{2\pi N} = \frac{p_s f a_p^2}{8}. \tag{3.28}$$

The force on the drill is more difficult to estimate. Assuming that the resultant forces act at the drill periphery, the force on each flute in the plane normal to the axis of rotation is:

$$F = \frac{P}{vN_t} = \frac{p_s v_f a_p}{4NN_t} = \frac{p_s f a_p}{4N_t} \tag{3.29}$$

Example

A standard 12 mm, 120° point angle, two-flute high speed steel (HSS) twist drills will be used to make 50 mm deep holes on aluminum workpiece. The maximum power available to drive the drills is 2 kW, the drilling time to produce one hole is not to be greater than 5 s. The transfer station must make 1200 holes before it can be stopped to change worn out drills. The tool maker specifies that the tangential force cannot exceed 0.75 kN or the drill may break. Empirically it is found that the tool life (L_t, in s) for a drill is governed by the cutting speed (v, in m/s) as $L_t/60 = (0.4/v)^{10}$. Determine the cutting conditions (feed rate and rotational speed) to analyze if it is possible to meet all the constraints on time, tool life, force, and power.

Solution

To meet the time limit, the feed rate should be at least

$$v_f = \frac{l_w + \frac{a_p}{2}\cot\kappa_r}{t} = \frac{50 + \frac{12}{2}\cot 60^\circ}{5} = 10.7 \text{ mm/s}.$$

To meet the tool change requirement, the tool life should be at least $5 \text{ s} \times 1200 = 6000 \text{ s}$, so the cutting speed should not exceed

$$v = \frac{0.4}{(6000/60)^{0.1}} = 0.25 \text{ m/s}.$$

Therefore, the spindle speed should not exceed $N = \frac{0.25 \times 10^3}{\pi(12)} = 6.63 \text{ rev/s}.$

The undeformed chip thickness (minimum expected) is

$$a_c = \frac{f}{N_t}\sin\kappa_r = \frac{v_f}{NN_t}\sin\kappa_r = \frac{10.7}{2 \times 6.63}\sin 60^\circ = 0.7 \text{ mm}.$$

The specific cutting energy is $p_s = \exp(-0.38 - 0.46\log(0.7)) = 0.5 \text{ GJ/m}^3 (\text{or J/mm}^3)$

The required power is $P = p_s \frac{\pi v_f a_p^2}{4} = (0.5)\frac{\pi(10.7)12^2}{4} = 605 \text{ W}$, which is less than 2 kW. OK.

The tangential force on each flute is $F = \frac{p_s f a_p}{4N_t} = \frac{(0.5)(10.7/6.63)(12)}{4(2)} = 1.21 \text{ kN}.$

Since this exceeds the breakage force limit of the tool, it is not possible to meet the life, time, and force requirements all at the same time. One solution to this problem is to "gang" drill the holes using a drill head with multiple spindles so that several holes can be made simultaneously at a reduced feed rate. It is interesting to note that a higher rotational speed saves the tool from breakage, but at the same time causes faster tool wear.

Example

A three-flute twist drill of 8 mm diameter and 62° major cutting-edge angle is used to make a 50 mm deep hole on a cast iron part. The drill press offers a spindle speed ranging from 120 to 5000 rpm. Find the minimum power (in W) needed to drill the hole in less than 45 s.

Solution

From Eq. (3.25), $t = \dfrac{l_w + \dfrac{a_p}{2}\cot \kappa_r}{fN} \Rightarrow \dfrac{45}{60} > \dfrac{50 + \dfrac{8}{2}\cot 62^\circ}{fN} \Rightarrow fN > 69.5$ mm/min.

Use the smallest spindle speed $N = 120$ rpm so as to maximize $f = 0.58$ mm/rev for the smallest possible specific cutting energy.

From Eq. (3.24), $a_c = \dfrac{f}{N_t}\sin \kappa_r = \dfrac{0.58}{3}\sin 62^\circ = 0.17$ mm, therefore, $p_s = 2.48$ J/mm^3 (Fig. 2.4).

From Eq. (3.27),

$$\text{the power required} = \frac{\pi fNa_p^2}{4} p_s = \frac{\pi(69.5)8^2}{4}(2.48) = 8{,}664 \text{ J/min} = 144 \text{ W}$$

Homework

1. Face milling is to be done on the heads of an engine using a 12-tooth cutter. In the following figure, the idealizations of the cutter and the engine head are shown. The specifics of this operation are: Cutting tool: 12 square carbide inserts with a 1 mm corner radius will be used. The insert holder and insert geometry give a radial and axial rake angle of 0° and a corner cutter angle of 60°. Workpiece: cast iron 500 mm long and 200 mm wide that will have 14 mm of the material face milled off the top. Cutting configuration: $N = 200$ rev/min and the chip load, $s_z = 0.5$ mm/tooth. Estimate the power (in kW) and torque (in Nm) required to machine this part. Estimate the time (in s) it will take to machine this part, and estimate the R_a roughness (in μm) of the surface.

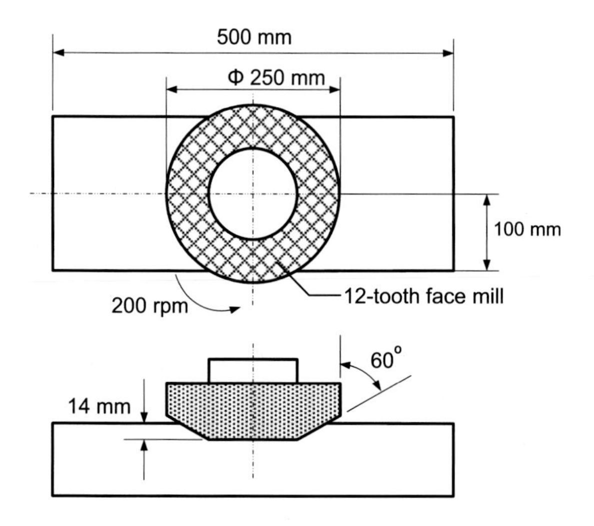

2. A carbon steel block is end milled by a machine of 1.5 kW maximum power. The process utilizes a four-flute cutter of 24.5 mm diameter at 5 mm radial depth of cut and 40 mm axial depth of cut in an up-cutting configuration. The length of workpiece being machined is 60 mm. If the total machining time is not to be more than 60 s, what would be the best surface finish achievable? Give your answer in terms of R_a. Since the depth of cut covers the entire width of the workpiece, the surface of concern is created along the wall of the cutter. Assume that the maximum cutting power is twice the average cutting power. There is no need to use safety factor of any sort in your calculation.
3. Peripheral end milling of cast iron workpiece with an inside corner, a straight section, and an outside corner are shown in the figure. This will be done on a CNC machine tool where the programmed feed rate is 4 mm/s and the rotational speed is programmed at $N = 180$ rev/min. (a) For the straight sections of this cut, estimate the average chip thickness, the MRR, and power; (b) repeat the above for the inside corner section; (c) repeat the above for the outside corner section; and (d) based on your calculations, what is an appropriate way to program the CNC machine tool path to keep the power consumption constant when making corners?

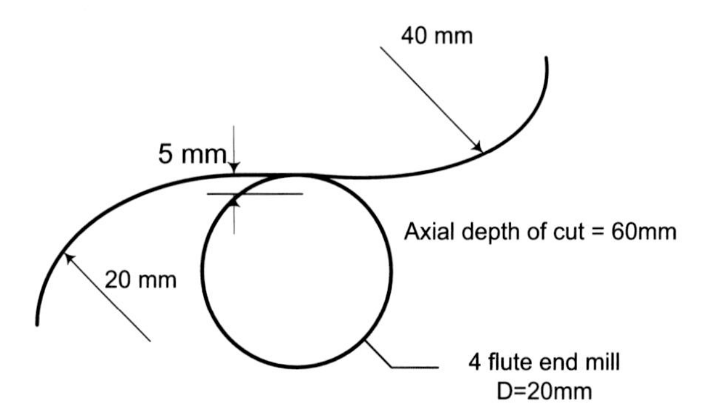

4. In drilling of a medium carbon steel workpiece with a 15 mm diameter HSS drill bit (point angle 150°, two flutes), it is observed that a down feed of 0.1 mm/rev, regardless of the spindle speed, can be applied safely without breaking the bit. If a 35 mm long hole is to be drilled within 20 s or less, what is the minimum drilling machine power (in J/s) required? Estimate the breaking torque of the drill bit (in J).
5. The figure shows face milling of a cast iron block with a four-blade, 100 mm diameter cutter at a radial depth of cut, a, of 50 mm, an axial depth of cut of 1 mm, a spindle speed of 1200 rpm, and a feed rate (v_f) of 50 mm/min. The cutting blades have corner angle of 15°, face cutting edge angle of 5°, and no significant nose radius. Estimate the resulting surface roughness (in μm) and the average machining power (in J/s). Note that for cast iron the specific cutting energy (p_s, GJ/m^3) is related to average chip thickness by (a_c, mm) $\log p_s = 0.04 - 0.46 \log a_c$.

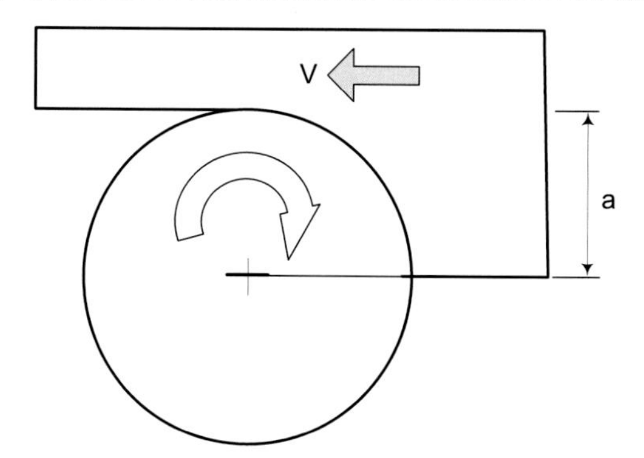

6. The understanding of the cutting force system plays an important part in the planning, monitoring, and controling of machining processes as well as in its traditional role in the design of tool geometry, machine structure, and servo mechanisms. The goal of this problem is to develop more scientific knowledge about multiflute cutting processes and their force responses.

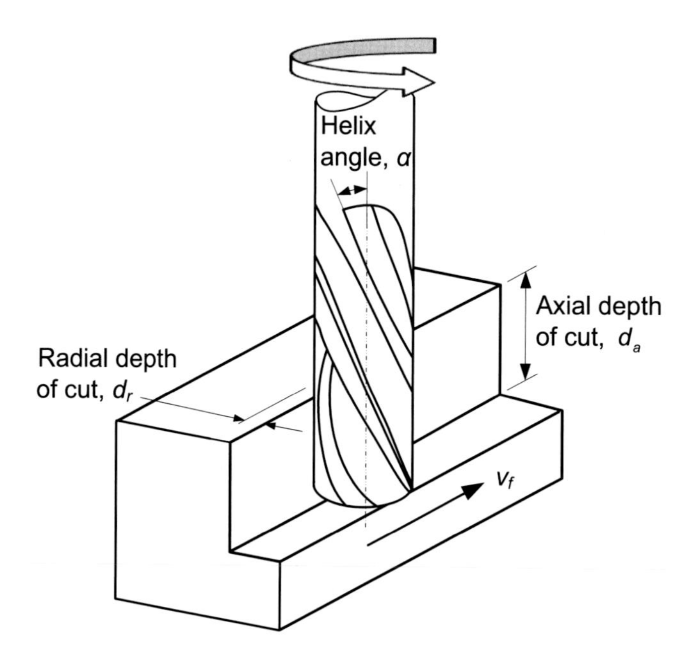

4 Grinding Processes

4.1 Grinding Machine and Process Overview

Grinding machines use abrasive wheels that rotate at high speed to either remove a layer of material (in rough grinding) or finish the part surface (in finish grinding). The process of grinding is much different from other cutting processes that a separate chapter of discussion is warranted. Two major differences between grinding and cutting (turning, milling, and drilling) are the scale of the chips produced and the amount of force/energy required. In light of the large number of cutting edges involved in grinding, the chips produced are usually quite small. It is difficult to estimate the undeformed chip thickness in grinding because the individual grains in the wheel are irregular. As a rough guide, chip thickness of 0.00025–0.025 mm can be assumed for a closely packed grain structure. Smaller grains, a higher cutting speed, a lower feed or feed rate, and a smaller depth of cut, all tend to result in a smaller chip thickness. Cutting processes have controlled geometry, usually with positive rake angles; however, grinding processes use small abrasive grits on the wheel with random orientation and geometry that provides rake angles of very negative values (−60° or larger). Due to the negative rake angles, rubbing and plowing actions between the wheel and the workpiece are brought about to increase the force and energy needed in removing the same amount of material in grinding than in cutting.

Although there are unfavorable aspects of grinding, fortunately they can be controlled by selecting the grinding conditions. For end milling, the axial depth of cut is on the order of 10 mm, while in grinding this is usually reduced by one order of magnitude at least. The wheel surface speed in grinding is often in the range of 10–25 m/s, while the corresponding cutting speed is often 10–20 % of this. When grinding conditions are selected correctly, several advantages of grinding can be utilized. First is part surface finish. In cutting, this is on the order of 1 μm R_a, as limited by the number of cutting edges. In grinding, the number of cutting edges increases by several orders of magnitude and the surface finish is usually several orders of magnitude better than that in cutting. Additionally, grinding is effectively used to pro-

S. Y. Liang, A. J. Shih, *Analysis of Machining and Machine Tools*,
DOI 10.1007/978-1-4899-7645-1_4

Table 4.1 Comparison of cutting and grinding processes

	Cutting edge geometry	Rake angle	Chip thickness (mm)	Specific cutting energy	Tool surface speed (m/s)	Surface roughness (μm R_a)	Type of work material
Cutting	Deterministic	+	>0.05	Low	<20	>1	Soft/ductile
Grinding	Stochastic	–	<0.025	High	>20	<1	Hard/brittle

cess hard and brittle materials such as tool material, ceramic components, or glass/optical parts. This is due to the small chip thickness generated that limits the extent of surface damage on the parts caused by brittle fracture (Table 4.1).

4.2 Grinding Wheel Code

The grinding wheel code (ANSI B74.13-1977 and ISO 525-1975E) is used to specify the abrasive type, its grit size, and the bonding method. For example, A54-K20V represents an aluminum oxide wheel, 54 grit, medium hardness, open structure, with a vitrified bond.

- The types of abrasives consist of a menu of (A) aluminum oxide, the most common abrasive for ferrous materials and is relatively soft and tough; (C) silicon carbide, the most common abrasive for nonferrous materials; (Z) zirconia alumina is one of the toughest abrasives and is used for high-impact rough grinding such as cutoff; (B) cubic boron nitride (CBN) is a superabrasive suitable for grinding ferrous materials, and is hard and brittle, typically bonded in a thin layer to a metallic hub due to its high cost; (D) diamond, another superabrasive that can be used for nonferrous materials and ceramics.
- Grit size is indicated by the number of wires per inch that the abrasives pass through. For example, 120 grit means that the abrasives pass through a screen with 120 wires per inch and do not pass through the next finest size of 150 grit. Therefore, the grit number is inversely proportional to the size, that is, 30 grit is larger than 36. Standard grit sizes are 8, 10, 12, 14, 16, 20, 24, 30, 36, 46, 54, 60, 70, 80, 90, 100, 120, 150, 180, 220, 240, 280, 320, 400, 500, 600, and so on.
- The grade or hardness is a letter code indicating how difficult it is to remove a grit from the wheel. It ranges from A to Z, with A indicating the lowest hardness.
- Structural openness is a numeric code (0–25) indicating the relative porosity or approximate spacing between grits with low numbers indicating tight packing.
- The type of bond used to hold the grits in the wheel matrix is indicated by a letter code. (V) is for vitrified bond, which is the most common inorganic bond. (B) designates a resinoid bond. (R) is rubber bond, which is the most resilient and toughest bond. (E) is used for the shellac bond, which is more heat sensitive than vitrified bonds. When temperature goes up slightly, the shellac bonded grits pull out to prevent workpiece from burning. (M) is for metal bonds, which are almost exclusively used for the superabrasives (CBN or diamond). M bonds are so strong that these expensive abrasives do not pull out.

4.3 Surface Grinding

The configuration of surface grinding is very similar to slab or peripheral millings. Surface grinding is performed to produce high tolerance, low surface roughness, and flat planar surface or channel. Figure 4.1 shows a surface grinder to create a flat surface in traverse grinding or to create a channel in plunge grinding. The material removal rate (MRR) in both cases is given by

$$Z_w = fa_p v_{trav}. \tag{4.1}$$

The stock entering the grinding zone and the material leaving as swarf is considered identical:

$$\begin{aligned} fa_p v_{trav} &= a_p h_{eff}(\pi D_s n_s \pm v_{trav}) \text{ in traverse} \\ fa_p v_{trav} &= fh_{eff}(\pi D_s n_s \pm v_{trav}) \text{ in plunge} \end{aligned}, \tag{4.2}$$

where the plus sign is for up grinding and minus for down grinding, n_s is the wheel spindle speed in rps and h_{eff} is defined as the effective chip thickness.

$$\begin{aligned} h_{eff} &= \frac{fv_{trav}}{(\pi D_s n_s \pm v_{trav})} \text{ in traverse} \\ h_{eff} &= \frac{a_p v_{trav}}{(\pi D_s n_s \pm v_{trav})} \text{ in plunge} \end{aligned}. \tag{4.3}$$

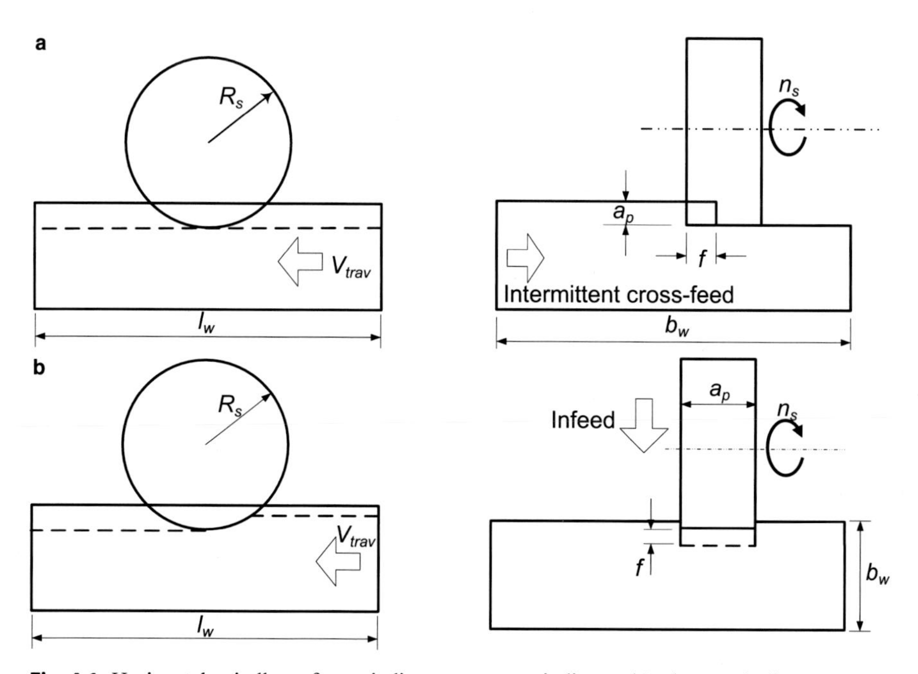

Fig. 4.1 Horizontal-spindle surface grinding **a** traverse grinding and **b** plunge grinding

While h_{eff} is not a physically measurable quantity, it is the size of an idealized continuous ribbon of material removed by grinding. It is an aggregate quantity that has been used to evaluate the specific grinding energy p_s. The power required at the grinding head is thus

$$P = p_s Z_w = p_s f a_p v_{trav}. \tag{4.4}$$

The grinding time t_m in surface grinding is given by

$$t_m = \frac{b_w}{2 f n_r}, \tag{4.5}$$

where b_w is the width of the workpiece in traverse grinding and the total depth of material to be removed in plunge grinding. The term n_r is the frequency of work full, back and forth, reciprocation is given by

$$n_r = \frac{v_{trav}}{2\left\{l_w + 2\sqrt{(D_s/2)^2 - (D_s/2 - a_p)^2}\right\}}. \tag{4.6}$$

Note that the extra grinding distance is allowed at the end of a stroke to provide room for the initial engagement in the following stroke. Also note that material is ground in both forward and backward strokes so the process is constantly switching between conditions of up- and down grinding.

One of the important measures of the overall performance of the grinding operation is the grinding ratio G. It is defined as the ratio of the workpiece MRR to the wheel wear rate

$$G = \frac{\text{volume of material removed}}{\text{volume of wheel wear}}. \tag{4.7}$$

The significance of the G-ratio can be seen in Fig. 4.2, where the MRR, power, and tool wear rate depend on the G-ratio. At the beginning of process, the minute amount of rubbing does not contribute to material removal. As the grinding force goes beyond certain threshold value, meaningful grinding starts to take place. Meanwhile, the tool holds up quite well when the grinding force stays below certain breakdown level and the mechanical and thermal loads are limited. When force greater than the breakdown level is applied, the wheel overheats, disintegrates, and wears fast. Therefore, there is a peak G-ratio right at the vicinity of the breakdown force level to represent the optimality of the process.

In the consideration of plunge surface grinding as an example, when the grinding wheel wears radially, that is, the radius decreases at a (linear) rate of v_r, the (volumetric) radial wheel wear rate is approximately

$$Z_r = \pi D_s a_p v_r. \tag{4.8}$$

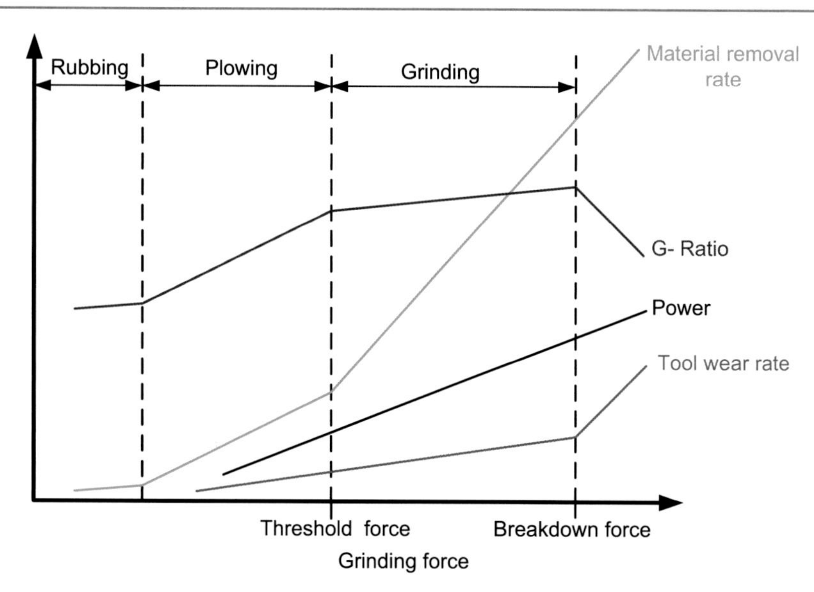

Fig. 4.2 The variation of MRR, power, and G-ratio relative to the force

Therefore the grinding ratio in the case of plunge surface grinding is

$$G = \frac{f v_{trav}}{\pi D_s v_r}. \tag{4.9}$$

The MRR largely depends on the process parameters, while the wheel wear rate depends on a number of things such as the tool characteristics, grinding fluid, and reactivity of the workpiece–abrasive combination. G-ratios on the order of 10–100 are common for conventional abrasives, while for superabrasives the values range from 100 to 1000. Based on the definition of wheel grade (designated by a letter ranging from A to Z), soft wheels tend to have lower G-ratios than wheels exhibiting hard wheel behavior.

In grinding, the plowing and rubbing interactions between the wheel and the workpiece result in progressive wear areas on the grains and in turn the increase of friction at the interface. Eventually, this friction becomes large enough either to tear the worn grain from the bond of the wheel and thus expose a new unworn grain or to fracture the worn grain to produce new cutting edges. Therefore, a grinding wheel has a self-sharpening characteristic, and the force a grain can withstand before being torn from the wheel or fractured is the most important factor when grinding wheel performance is considered.

A wheel consisting of relatively tough grains strongly bonded together will only exhibit the self-sharpening characteristic to a limited degree and will quickly develop a glazed appearance during grinding due to the large worn areas developed on the grains. These worn areas result in excessive friction, large force consumption, and quick overheating of the workpiece. Grinding under these inefficient conditions necessitates "dressing" the wheel with an ultra-hard material such as a diamond tip, which can remove or fracture the worn grains. Upon the exposure of sharp new cutting edges, the G-ratio is high, but it decreases quickly to a constant value. The steady state G-ratio typically lasts for an extended period of time where the most productive grinding process is done before it begins to drop again quickly due to accelerated wheel loading that justifies another dressing cycle.

The opposite characteristic, where the grains in the wheel are torn out easily, describes a soft wheel. In extreme cases, the volume of metal removed from a soft wheel can be significant compared to the volume of workpiece material removed; the wheel will rapidly lose its shape thereby necessitating dressing. Thus, the objective of selecting a wheel for a particular operation is to achieve a compromise between the two extremes of hard and soft wheels. In general, the following guidelines can be used for the selection of a grinding wheel: choose a hard-grade wheel for soft materials to delay wear and a soft-grade wheel for hard materials to facilitate self-sharpening; choose a large grit (hard wheel) for soft and ductile materials and a small grit (soft wheel) for hard and brittle materials; and choose a small grit for a good finish and a large grit for a maximum metal removal rate.

In addition to grain and bonding properties, grinding conditions also play an important part in determining whether a wheel will behave hard or soft. Figure 4.3 shows the approximate geometry of a layer of material removed by a single grain during plunge grinding. The average length of a chip l_c during grinding can be approximated by

$$l_c = \frac{D_s}{2}\theta \approx \frac{D_s}{2}\sin\theta, \tag{4.10}$$

where D_s is the diameter of the wheel and

$$\cos\theta = \frac{(D_s/2) - f}{D_s/2} = 1 - \frac{2f}{D_s} \tag{4.11}$$

with

$$\sin\theta = \sqrt{1-\cos^2\theta} = \sqrt{\frac{4f}{D_s} - \frac{4f^2}{D_s^2}}. \tag{4.12}$$

Substituting the above in Eq. (4.10) and neglecting the second-order term gives

$$l_c = \sqrt{fD_s - f^2} \approx \sqrt{fD_s}. \tag{4.13}$$

The chip produced can have random geometry depending on the shape of the grain. Suppose the chips have the triangular cross section shown in Fig. 4.3, the average volume V_o of each chip is given by

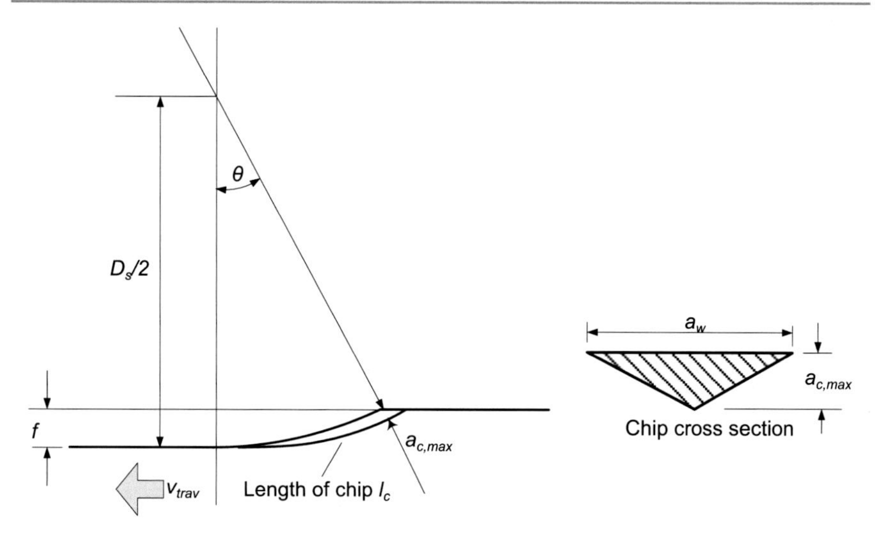

Fig. 4.3 Geometry of chip removal in plunge grinding

$$V_o = \frac{1}{4} a_w a_{c,max} l_c, \tag{4.14}$$

where a_w is the average width of the chip and $a_{c,max}$ is the maximum undeformed chip thickness, and they can be related by a grain aspect ratio r_g as

$$r_g = \frac{a_w}{a_{c,max}}. \tag{4.15}$$

The number of chips produced per unit time N_c is given by

$$N_c = v_t a_p C_g, \tag{4.16}$$

where v_t is the surface speed of the wheel and C_g is the number of actively cutting grains per unit area on the wheel surface. A common method to estimate C_g is

$$C_g = \frac{1}{10} \left(\frac{1}{\frac{\pi d_g^2}{4}} \right) \approx \frac{1}{10 d_g^2}, \tag{4.17}$$

where d_g is the average grit diameter which is often agreed upon as 60% of the wire spacing that determines the grit size. For example, a 120 grit passes through a 1/120 in. wire spacing and it has 0.005 in. average grit diameter. The above equation stipulates that only one in every 10 grits is actually cutting.

Since $Z_w = V_o N_c$ from Eqs. (4.1) and (4.13) to (4.16)

$$a_{c,max}^2 = \frac{4v_{trav}}{C_g r_g v_t}\sqrt{\frac{f}{D_s}}. \tag{4.18}$$

Since $a_{c,max}$ indicates the extent of grain penetration into the workpiece, a larger $a_{c,max}$ will result in greater force acting on each grain during grinding. It follows that any change in grinding conditions tending to increase $a_{c,max}$ increases the self-sharpening process, and the grinding wheel tends to appear softer. Thus, from the above equation the following changes in grinding conditions would be expected to make a wheel behave softer: an increase in the work traverse speed v_{trav}, an increase in the infeed f, and a decrease in the wheel surface speed v_t. Note that Eq. (4.18) is derived based on the configuration of plunge grinding. The infeed f should be substituted by a_p for traverse grinding.

Example

A stainless steel was plunge ground with an 8 in. diameter silicon carbide bonded abrasive wheel of fine grain and medium density. The infeed used was 0.3×10^3 in. and the work traverse speed was 1.25 ipm, and it was felt that the wheel hardness in this application is appropriate. Suppose a 6 in. wheel of the same grain size and density is to be used and the new hardness should not be greater than the previous in accomplishing the same MRR, what should be the limitations on the new infeed and the new traverse speed?

Solution

$$a_{c,max}^2 = \frac{4v_{trav}}{C_g r_g v_t}\sqrt{\frac{f}{D_s}} = \frac{4Z_w}{\sqrt{fD_s}\,a_p C_g r_g v_t}\left(\text{since } v_{trav} = \frac{Z_w}{fa_p}\right)$$

$$= \frac{4Z_w}{\sqrt{fD_s^3}\,a_p C_g r_g \pi n_t}\,(\text{since } \mathrm{v_t} = \pi D_s n_t),$$

where n_t is the wheel rotational frequency. To have a larger new $a_{c,max}$ than the previous,

$$f'D_s'^3 < fD_s^3 \Rightarrow f < \frac{fD_s^3}{D_s'^3} = \frac{(0.3)8^3}{6^3} = 0.7 \times 10^{-3} \text{ in.}$$

Also, for the same Z_w, the infeed f should be inversely proportional to v_{trav}, thus

$$\frac{D_s'^3}{v'_{trav}} < \frac{D_s^3}{v_{trav}} \Rightarrow v'_{trav} > v_{trav}\frac{D_s'^3}{D_s^3} = \frac{(1.25)6^3}{8^3} = 0.527 \text{ ipm.}$$

A way to estimate the achievable ground surface finish is to use the analogy between end milling and surface grinding. In end milling, it is known that

$$R_t \approx \frac{s_z^2}{8\left(\frac{D}{2} \pm \frac{s_z N_t}{\pi}\right)}. \tag{3.19}$$

The wheel diameter D_s is used instead of the milling cutter diameter D. The $(S_z N_t / 2\pi)$ term is often ignored since it is way too small in comparison to $(D/2)$ in grinding. The counterpart of s_z in surface grinding is the feed per effective grit given by

$$s_z = \frac{v_{trav}}{v_t} L, \tag{4.19}$$

where L is the spacing between successive cutting points, and thus L/v_t is the time interval between cutting. The L is difficult to measure directly but can be estimated by comparing Eq. (4.18) and the geometrical definition of maximum chip thickness in Fig. (4.3)

$$a_{c,max} = s_z \sin\theta \tag{4.20}$$

such that

$$L \approx \frac{2}{C_g a_w} = \frac{2}{C_g r_g a_{c,max}}. \tag{4.21}$$

Example
A 4340 steel plate of dimensions 200×300 mm (along the traverse speed direction) by 50 mm is to be surface ground to a depth of 0.1 mm by a 150 mm diameter 60 grit grinding wheel. The wheel speed is 20 rps and each workpiece reciprocation cycle takes 4 s with a cross feed of 2 mm. Estimate the surface roughness and the required power at wheel spindle.

Solution
From Eq. (4.6),

$$v_{trav} = n_r \left\{ 2\left(L + 2\sqrt{\left(\frac{D_s}{2}\right)^2 - \left(\frac{D_s}{2} - a_p\right)^2} \right) \right\}$$

$$= \frac{1}{4}(2)\left(300 + 2\sqrt{\left(\frac{150}{2}\right)^2 - \left(\frac{150}{2} - 0.1\right)^2}\right) = 154 \text{ mm/s}$$

$$d_g = (0.6)(1/60)(25.4) = 0.254 \text{ mm.}$$

The number of actively cutting grains per unit area (mm^2)

$$C_g \approx \frac{1}{10 d_g^2} = 1.55 \, .$$

The maximum undeformed chip thickness

$$a_{c,max}^2 = \frac{4 v_{trav}}{C_g r_g v_t} \sqrt{\frac{a_p}{D_s}} = \frac{4(154)}{(1.55)(10)(150\pi)(20)} \sqrt{\frac{0.1}{150}} \Rightarrow a_{c,max} = 0.010 \text{ mm.}$$

From Eq. (4.19) $\Rightarrow s_z = \dfrac{2 v_{trav}}{v_t C_g r_g a_{c,max}} = \dfrac{(2)(154)}{\pi(150)(20)(1.55)(10)(0.01)} = 0.211$ mm.

From Eq. (3.19) in Chap. 3 $\Rightarrow R_t \approx \dfrac{0.211^2}{8(150/2)} = 7.41 \times 10^{-5}$ mm.

From Eq. (4.3) $\Rightarrow h_{eff} = \dfrac{(2)(154)}{\pi(150)(20)} = 0.033$ mm, note that ± 154 mm/s is negligible in this case.

Figure 2.4 $\Rightarrow p_s = 10^{0.19 - 0.46 \log 0.033} = 7.44$ J/mm^3.

Equation (4.4) $\Rightarrow P = (7.44)(2)(0.1)(154) = 229$ W.

From the example it is seen that grinding is generally a high finish, high specific energy, and low power process, in comparison to turning and milling processes.

4.4 Cylindrical Grinding

Cylindrical grinding is also referred to as chuck grinding, and it involves the rotation of grinding wheel as well as the workpiece. This type of grinding, along with centerless grinding which is to be discussed later, is the most common finishing operation used to make bearings and bearing raceways. These processes are similar to cylindrical turning, and grinding attachments are available that allow these operations to be performed on an engine lathe. In addition to external surfaces, internal surfaces can also be ground based on similar principles.

Generally speaking, there are two types of cylindrical grinding, namely, traverse and plunge. Figure 4.4a and b show the respective configurations. In traverse grinding, the axial feed rate v_{trav} is given by

$$v_{trav} = f n_w, \tag{4.22}$$

where f is the feed per work revolution and n_w is the work rotational speed in rpm. The MRR is

$$Z_w = \pi a_p D_w v_{trav}. \tag{4.23}$$

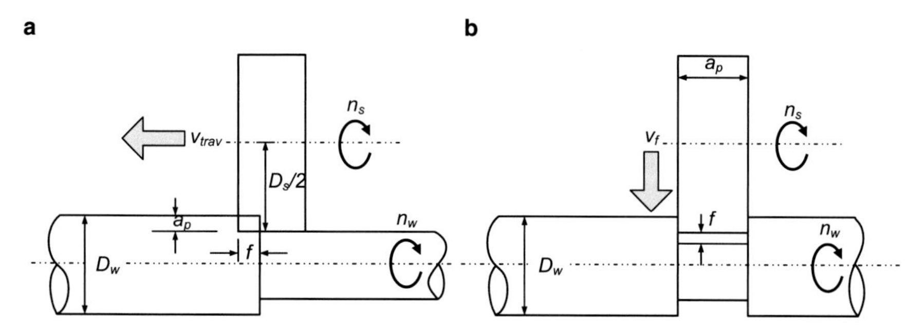

Fig. 4.4 Cylindrical-grinding operations. **a** Traverse grinding and **b** plunge grinding

In plunge grinding, the wheel is fed into the workpiece without traverse motion being applied, to form a groove. Its MRR decreases as the wheel feeds deeper into the workpiece. The maximum metal removal rate is given by

$$Z_{w,max} = \pi v_f D_w a_p, \tag{4.24}$$

where v_f is the linear velocity of the wheel feed motion that can be expressed as

$$v_f = f n_w. \tag{4.25}$$

The concept of effective chip thickness in cylindrical grinding can be extended from that in surface grinding. In traverse grinding, consider that the stock entering the grinding zone and the material leaving as swarf are identical,

$$f a_p \pi D_w n_w = a_p h_{eff} \pi (D_s n_s \pm D_w n_w); + \text{ for up and } - \text{ for down grinding} \tag{4.26}$$

which suggests that

$$h_{eff} = f \frac{D_w n_w}{D_s n_s \pm D_w n_w}, \tag{4.27}$$

and in plunge cylindrical grinding the definitions of a_p and f are different than they are in Eq. (4.26); thus, the formula to calculate h_{eff} is identical to Eq. (4.27) with f replaced by a_p.

The power required at grinding wheel spindle can then be estimated from

$$\begin{aligned} P &= p_s Z_w = p_s \pi v_{trav} D_w a_p \text{ in traverse grinding} \\ P &= p_s Z_w = p_s \pi v_f D_w a_p \text{ in plunge grinding} \end{aligned} . \tag{4.28}$$

The torque on the wheel T_s and the tangential force F_t in plunge grinding are

$$T_s = \frac{P}{2\pi n_s} = \frac{p_s v_f D_w a_p}{2n_s}$$
$$F_t = \frac{P}{v_s} = \frac{p_s v_f D_w a_p}{D_s n_s}$$ (4.29)

The normal force F_n can be estimated based on an empirical force ratio coefficient c_g as

$$F_n = \frac{F_t}{c_g} = \frac{p_s v_f D_w a_p}{c_g D_s n_s} = \frac{p_s (f n_w) D_w a_p}{c_g D_s n_s}. \quad (4.30)$$

The value for c_g depends on the wheel loading condition. Approximate values are 0.7 for sharp or freshly dressed wheels, 0.5 for constant *G*-ratio wheel condition, and 0.3 for dull or loaded wheels.

In plunge grinding operation, where the wheel is fed in a direction normal to the work surface (infeed), the feed will initially be less than the nominal feed setting on the machine. This feed differential is a result of the deflection of the machine tool elements and workpiece under the forces generated during the operation. Thus, on completion of the theoretical number of revolutions required, some work material will still have to be removed. The operation of removing this material is referred to as the sparking-out, and it is achieved by continuing the original grinding operation with no further application of feed until metal removal becomes insignificant (no further sparks appear). The machining time given by Eq. (4.5), therefore, underestimates the actual value by the time for sparking-out in surface grinding. For the cylindrical grinding processes the grinding force cycle is as shown in Fig. 4.5. The spark-out time, t_s, needs to be carefully planned in the grinding cycle; otherwise, the tolerance and finish of parts may not meet technical specifications.

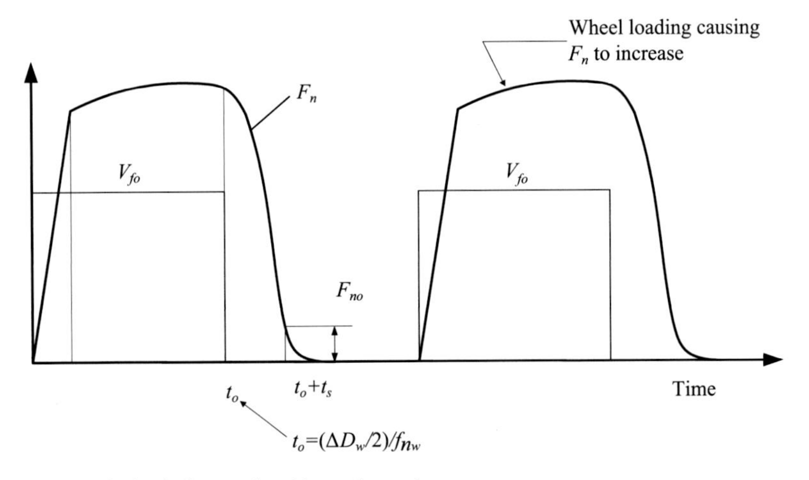

Fig. 4.5 A typical grinding cycle with spark-out time

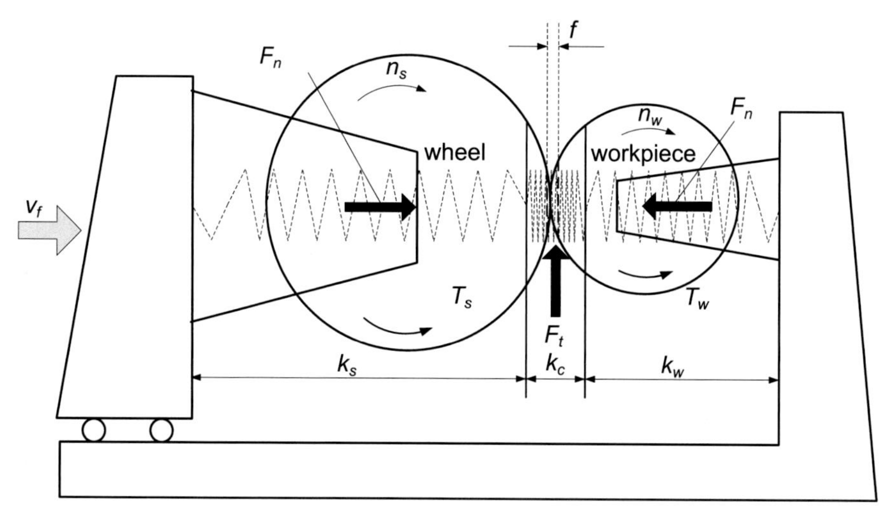

Fig. 4.6 Force analysis in cylindrical plunge centerless grinding

In Fig. 4.6, the grinding system stiffness has an overall stiffness k_{eff} that can be considered as the serial superposition of the workpiece stiffness k_w, the wheel and machine stiffness k_s, and the contact stiffness k_c at the interface.

$$k_{eq} = \left(\frac{1}{k_s} + \frac{1}{k_c} + \frac{1}{k_w} \right)^{-1}. \tag{4.31}$$

The normal grinding force causes the grinding system to deflect in the radial direction by an amount governed by the stiffness as

$$\delta = \frac{F_n}{k_{eq}} \tag{4.32}$$

From Eq. (4.30), it is understood that the normal force is proportional to the feed per revolution. By defining a proportionality constant (referred to as the grinding stiffness) $k_g = \dfrac{p_s D_w a_p n_w}{c_g D_s n_s}$,

$$F_n = k_g f. \tag{4.33}$$

Therefore,

$$f = \delta \frac{k_{eq}}{k_g}. \tag{4.34}$$

In the case of changing normal force, the system deflection shows a rate of change that alters the instantaneous plunging speed $v_f(t)$

$$v_f(t) - \frac{d}{dt}\delta(t) = f(t)n_w. \tag{4.35}$$

Considering the spark-out cycle given in Fig. 4.5, upon the termination of plunging feed motion $v_f(t)$, the grinding force exhibits a decreasing free response. Substituting $\tau = t - t_o$, $v_f(\tau) = 0$, for $\tau > 0$; and $f(\tau) = \delta(\tau)\frac{k_{eq}}{k_g}$, into the above equation gives

$$\frac{d\delta(\tau)}{dt} = -n_w \frac{k_{eq}}{k_g}\delta(\tau), \tag{4.36}$$

for which the initial condition is that at $\tau = 0$, $\delta(0) = f(0)\frac{k_{eq}}{k_g} = \frac{v_f(0)}{n_w}\frac{k_{eq}}{k_g} = \frac{v_{fo}}{n_w}\frac{k_{eq}}{k_g}$.

The solution for Eq. (4.36) is thus

$$\delta(\tau) = \frac{v_{fo}k_g}{n_w k_{eq}} e^{\left(\frac{-n_w k_{eq}}{k_g}\right)\tau}. \tag{4.37}$$

Alternatively, from Eq. (4.32)

$$F_n(\tau) = \frac{v_{fo}k_g}{n_w} e^{\left(\frac{-n_w k_{eq}}{k_g}\right)\tau}. \tag{4.38}$$

The determination of spark-out time has to be based on a certain criterion with respect to deflection or force. A threshold force, F_{no}, is often used for this purpose since normal forces below F_{no} do not remove material. So the spark-out time t_s can be defined as the time it takes from the termination of feed to the disappearing of force to the F_{no} level:

$$t_s = \frac{k_g}{k_{eq}n_w}\ln\left(\frac{v_{fo}k_g}{F_{no}n_w}\right). \tag{4.39}$$

Thus, the spark-out time depends on many variables, including not only machine characteristics but also process parameters and tool/work geometries. In general, the reduction of spark-out time requires a lower grinding stiffness k_g, so that the normal grinding force is lower for the same amount of infeed. It also requires a higher equivalent system stiffness k_{eq} (dominated by the most compliant element in the system) and a higher work speed n_w. Note that the above analysis assumes that the specific grinding energy does not vary with the deflection. If the variation of specific grinding energy is factored in, the required amount of spark-out time is expected to be greater than Eq. (4.39), because as the $f(\tau)$ gets smaller the specific grinding energy increases and so does the grinding stiffness.

Example
The outer diameter of a bearing raceway (OD=50 mm) is cylindrically (down) plunge ground by a wheel of 65 mm diameter on a grinder that has 5 kW power available at the spindle. The width of grind is 12 mm, spindle speed is 30 rps, the workpiece rotational speed is 2.5 rps, and the part roundness tolerance is 0.1 μm. The workpiece material is hardened 52100 steel. The contact, machine, and workpiece stiffness are 100, 140, and 800 kN/mm, respectively. Determine the radial feed per workpiece revolution to use the available spindle power completely. Also estimate the total grinding cycle time including grinding and spark-out for a grinding allowance of 0.2 mm. Assume that the force ratio coefficient c_g is 0.5.

Solution
Equation (4.27)$\Rightarrow$ $h_{eff} = f\dfrac{D_w n_w}{D_s n_s - D_w n_w} = \dfrac{f(50)(2.5)}{(65)(30)-(50)(2.5)} = 0.068f$ mm

From Fig. 2.4, $p_s = 10^{0.19-0.46\log(0.068f)}$
Equation (4.28)$\Rightarrow$ $5000 = 10^{0.19-0.46\log(0.068f)}\pi(50)(12)(2.5)f$

$$\Rightarrow f = 0.05 \text{ mm/rev}$$

Therefore,

$$k_g = \frac{p_s D_w a_p n_w}{c_g D_s n_s} = \frac{10^{0.19-0.46\log(0.068(0.05))}(50)(12)(2.5)}{(0.5)(65)(30)} = 32.5 \text{ kN/mm}$$

Equation (4.31) $\Rightarrow k_{eq} = \left(\dfrac{1}{100 \text{ kN/mm}} + \dfrac{1}{140 \text{ kN/mm}} + \dfrac{1}{800 \text{ kN/mm}}\right)^{-1} = 54$ kN/mm

For a part tolerance of 0.0001 mm, the threshold force is
$F_{no} = \delta k_{eq} = (0.0001)(54) = 0.0054$ kN

Equation (4.39)

$$\Rightarrow t_s = \frac{k_g}{k_{eq} n_w}\ln\left(\frac{v_{fo} k_g}{F_{no} n_w}\right) = \frac{k_g}{k_{eq} n_w}\ln\left(\frac{f\,k_g}{F_{no}}\right) = \frac{32.5}{(54)(2.5)}\ln\left(\frac{(0.05)(32.5)}{0.0054}\right) = 1.37 \text{ s}$$

$$\left(\text{Note that } v_{f0} = f(0)n_w, \text{ thus } t_s = \frac{k_g}{k_{eq} n_w}\ln\left(\frac{f(0)k_g}{F_{no}}\right)\right)$$

$$t_o = \frac{0.2}{0.05(2.5)} = 1.6 \text{ s}$$

The total cycle time is $t = 1.6 + 1.37 = 2.97$ s

4.5 Special Grinding Processes

4.5.1 Creep-Feed Grinding

The basic setup of creep-feed grinding is similar to that of surface grinding, although cylindrical creep-feed grinding is also possible, except for the depth of grind and the traverse speed. In conventional surface grinding (also called pendulum grinding), the depth of grind is typically very small and the traverse speed is high. The wheel is reciprocated back and forth across the workpiece, removing only a thin layer with each pass. In creep-feed grinding (also called swing grinding or deep grinding), all the material is removed in a single pass of the wheel with large depth of grind and relatively low traverse speed. In this manner the required grinding time can be significantly shortened. It is often applied to materials that are difficult to grind such as hydraulic pump slots and gas turbine blade base tree pattern.

The following table shows the typical differences between pendulum and deep grindings. Since deep grinding is performed at a depth 1000 times of the pendulum grinding and at a traverse speed 1/100 of pendulum grinding, it is expected to accomplish an overall MRR 10 times higher. It also involves longer thin chips, higher forces, but lower levels of force per chip—due to a longer contact length and a larger contact area. The lower force per chip suggests that softer wheels can be used. However, creep-feed grinding often brings about higher temperature that makes wheel life and surface integrity problematic. It is therefore advisable to use lower wheel speed with effective cooling. A wheel with more openness in the structure is also advantageous to provide grinding fluid reservoirs.

Because of the higher forces involved, machines for creep-feed grinding demand greater rigidity, stability, and power (3–5 times) than machines used for conventional surface grinding. It also requires continuous dressing such that the abrasive particles removed get washed clear to keep the temperature down and the specific grinding energy low (Table 4.2).

4.5.2 Centerless Grinding

Centerless grinding is very similar to cylindrical grinding except for the manner by which the workpiece is held and driven. In cylindrical grinding, the workpiece is held and driven by a chuck connecting to the headstock. In centerless grinding, the workpiece is supported by a work blade and is friction driven by a regulating wheel (typically rubber bond to give good friction) as seen in Fig. 4.7. The weight of the workpiece, the vertical grinding force, and the vertical friction provide the necessary contact between the workpiece and the work blade.

Similar to cylindrical or surface grinding, centerless grinding can be performed in either plunge or traverse mode. In the latter case, the process is referred to as through-feed centerless grinding, since the workpiece is fed axially through the two wheels. This is achieved by setting the axis of the regulating wheel to a small angle (a few degrees) to the grinding wheel axis. This causes the regulating wheel

Table 4.2 Comparison between conventional and creep-feed grinding setups

Condition	Conventional surface grinding	Creep-feed grinding
Wheel	A60 H8V	Same
Diameter (mm)	254	Same
Surface speed (m/s)	30.5	Same
Depth of grind (μm)	5.1	5100
Traverse speed (m/s)	0.31	0.0031
Active grit per area, C_g (cm^{-2})	186	93
Grain aspect ratio (r_g)	20	30
Wheel-work contact length (mm)	1.14	11.4
Tangential force/width (N/cm)	28	84.1
Radial force/width (N/cm)	56	168.1
Tangential force/grit (N)	1.33	0.8
Radial force/grit (N)	2.66	1.6

to provide not only rotation of the workpiece but also axial feed motion. In plunge centerless grinding, the grinding and the regulating wheel axes are perfectly parallel and the regulating wheel is fed inward at a fixed rate.

The centerless grinding process is widely used in the bearing, automotive, and aerospace industries. It is well suited to the long, slender workpiece where deflection is a problem in conventional cylindrical grinding. In centerless grinding the workpiece finds its own center and is completely supported in the grinding zone.

Since the supported surface (between workpiece and work blade) and the ground surface (between workpiece and grinding wheel) are the same, there is a tendency for feedback leading to the development of lobes on the ground surface. This lobbing effect is a geometric instability that typically takes place if the following condition is met:

$$\alpha = a\beta \text{ and } \pi - \beta = b\beta \text{ (both } \alpha \text{ and } \beta \text{ in radian)}, \tag{4.40}$$

where a is an even integer and b is an odd integer. Conventionally, β is set at 7° or 8°.

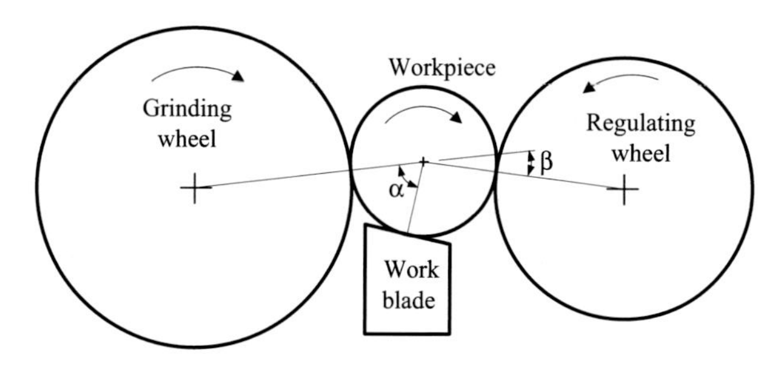

Fig. 4.7 A centerless grinding operation

In addition to geometric instability, there is dynamic instability that also affects the rounding ability. To avoid dynamic instability, a high-stiffness machine is required.

Homework

1. A 75 mm diameter, 15 mm wide A80-M12V grinding wheel is used for traverse surface grinding of a 50×50 mm 1040 steel workpiece. The workpiece traverse speed is 1 mm/s, the radial depth of grind is 1.5 mm, the (axial) intermittent cross feed is 10 mm/stroke, and the wheel surface speed is 314 mm/s. Estimate the required amount of grinding time (without spark-out) and the spindle power.
2. Develop the governing equation and the solution for spark-out time in a surface plunge grinding operation. Consider two spark-out approaches (1) to achieve the dimensional tolerance by one stroke of spark-out cycle with reduced wheel feed rate, or (2) to achieve the dimensional tolerance by multiple strokes of spark-out cycles with the same wheel feed rate as the grinding cycle.
3. In cylindrical traverse grinding, a 50 mm diameter steel bar is ground to 49 mm using a 100 mm diameter Al_2O_3 wheel. The wheel speed is 2400 rpm and the work speed is 1200 rpm while the total length of grind is 120 mm. It is estimated that the overall grinding system stiffness is 5 kN/mm and the force ratio coefficient is roughly 0.5. After a grinding pass is done at an axial feed of 0.05 mm per work revolution, a spark-out pass is exercised (may be under an axial feed different from 0.05 mm) to bring the diameter tolerance to within 0.002 mm. What is the least amount of time required to complete this spark-out pass? You may assume that the specific grinding energy is always 5 J/mm^3 (fixed) for this material.
4. Derive Eq. (4.40). Note that the derivation can be quite involved; however, it has been documented in research papers. Review of literature is recommended for this problem. References: Y. Furukawa, M. Miyashita, S. Shiozaki, Vibration analysis and work-rounding mechanism in centerless grinding, International Journal of Machine Tool Design and Research 11 (1971) 145–175 and W.B. Rowe, M. Miyashita, W. Koenig, Centerless grinding research and its application in advanced manufacturing technology, Annals of CIRP, 38/2 (1989) 617–625.

5 Machine Tool Components

The power, precision, accuracy, and speed of a machine tool largely affect the quality and cost of the produced parts and products. Designers have to consider latitude of factors including dimensions, material, configuration, and power source of the machines in order to deliver the functionality as demanded by the users. Advances in machine tool design and fabrication philosophy are quickly eliminating the differences among machine types. Fifty years ago, most machine tools had a single function such as drilling or turning, and operated alone. With the addition of automatic turrets, tool changers, and computer numerical control (CNC) systems, lathes become turning centers and milling machines become machining centers. Turning centers can also become machining centers with the addition of live or powered tool spindles in addition to the traditional single-point tools. These multiprocess centers can perform all the standard machining functions: turning, milling, boring, drilling, and even grinding. In this chapter, we will examine the machine tool structure from the standpoint of basic components, which are the frames, ways, motors, spindles, tools, and control units. The characteristics of these components cut across the machine lines and are fundamentally important to the operation and performance of general machines.

In the past, precision of machined part depended chiefly on the operator's capability. Today, accuracy is more dependent on the quality of the machine elements and resolution of the control. Conventional lathes routinely offer an accuracy of 3–5 μm at stable temperature condition. The ones that can achieve tolerances less than 1 μm are considered precision machines while those that achieve tolerances less than 0.1 μm should be considered ultraprecision machine tools.

S. Y. Liang, A. J. Shih, *Analysis of Machining and Machine Tools*,
DOI 10.1007/978-1-4899-7645-1_5

5.1 Frames

Frame is a machine tool's most basic element. It carries all the active and passive components and subassemblies spindles, slides, table, drives, and controls. The form of frames is basically dependent upon the position, length, and orientation of the moving axes, and the spatial arrangement of the components and subassemblies. The frame design is also influenced by the magnitude of the process forces, the thermal and environmental conditions, and accessibility for their own construction, as well as their use and operation.

Factors governing the choice of frame materials are: resistance to deformation (hardness), resistance to impact and fracture (toughness), limited expansion under heat (coefficient of thermal expansion), high absorption of vibrations (damping), and resistance to shop-floor environment (corrosion resistance), and low cost. Most frames are made of the following materials:

- Cast iron, a traditional material for machine tool frames, has good stiffness (modulus of elasticity $E = 50-110$ GN/m^2), strength ($\sigma_u = 100-300$ MN/m^2), and damping qualities. Geometrical features difficult to obtain any other way can be cast in and the cost is often low with large quantity. On the down side, casting size is limited by the expense of patterns, the problems with bolted joints, and the need to anneal large sections.
- Welded steel has a higher modulus ($E = 210$ GN/m^2) and strength ($\sigma_u = 400-1300$ MN/m^2), and is usually ribbed to provide stiffness. With welding, it is easy to make large sections and to add features even after the initial design. The main concern for welded steel is heat distortion, and circulating coolant through the structure or adding lead or sand to frame cavities is recommended.
- Composites, including those with polymer, metal, and ceramic matrices, may change machine tool design dramatically. Both matrix and reinforcing material can be tailored to provide strength in specific axes. Composite frames are usually expensive and have not yet generated a track record in the industry, but they have given great potential for high-speed, high-accuracy applications. One important design factor is the consideration of different expansion coefficients between the composite and the metal sections to which it is joined.
- Ceramics were introduced by the Japanese as a frame material in the 1980s. Ceramics offer strength, stiffness, dimensional stability, corrosion resistance, and excellent surface finish, but they are brittle and expensive.
- Reinforced concrete in the conventional form provides mass and reduces vibration. Another form, actually a polymer matrix composite made of crushed concrete or granite bound in a polymer matrix, is more popular. It was introduced by a Swiss company and has better damping characteristics than cast iron, can be cast into almost any shape, needs no stress relieving, and can accommodate fasteners and rails if inserts are used. However, it is not as strong ($\sigma_u = 5-60$ MN/m^2) and stiff ($E = 20$ GN/m^2) as metals and diffuses heat less efficiently.

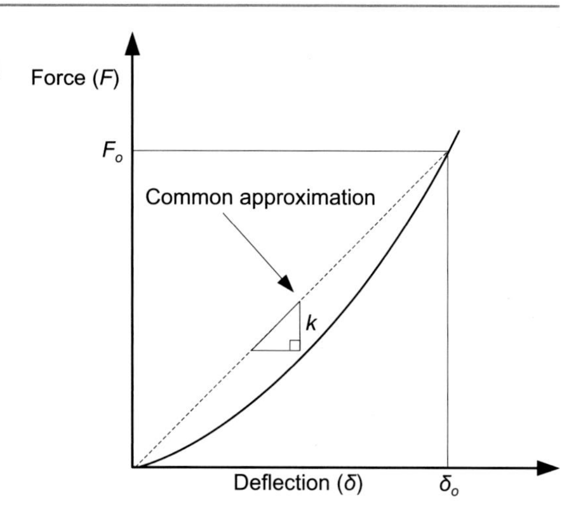

Fig. 5.1 The force versus deflection of a machine tool structure and approximated stiffness

The consideration of frame stiffness and damping is of particular importance to the overall functionality of machine tools. This is related to the machine's capability to resist static and dynamic forces during cutting. The static forces, along with the mass of the machine tool and the workpiece being machined, produce deflections and distortions which introduce geometric errors on the final product. The dynamic forces are results of intermittent cutter engagement and regenerative chip thickness variation. They can lead to forced vibrations or chatter which greatly lower the achievable product quality and tool life.

The static characteristic of a machine tool is dependent upon the elastic deformations in response to the applied loads at any given time. To complicate the issue is the fact that applied loads can vary as cutting progresses, thus the resultant deformation does not remain constant. In general, at any given time, the relationship between a deformation δ and the applied load F shows disproportional increase due to the increase of contact areas between machine elements. Figure 5.1 illustrates such a characteristic trend. It is a common practice to assume a linear relationship (which can only be found on jointless machines) and use an average stiffness, k, for simple calculations.

The deformations seen at the cutting point are the result of all force transmitted from all relevant machine components and constructional units. The static stiffness of the overall machine can often be considered as the elastic combination of the static stiffness of all components involved based on the analogy of spring in series or in parallel connections. Figure 5.2 shows a typical stiffness analysis of a horizontal boring machine with load $F_x = F_y = F_z = 40{,}000$ N acting in the respective coordinate axes on the spindle when it is in the working position.

Example
The figure shows a vertical machining center with a tool, a workpiece, and a fixture plan. The stiffness of each component has been identified as listed. Estimate the stiffness of the total machine tool system.

Vertical stiffness :
k_1 = tool, 45 kN/mm
k_2 = spindle & housing, 20 kN/mm
k_3 = spindle head, 62 kN/mm
k_4 = frame, 25 kN/mm
k_5 = guide ways, 78 kN/mm
k_6 = table, 120 kN/mm
k_7 = fixture A, 35 kN/mm
k_8 = fixture B, 42 kN/mm
k_9 = workpiece, 58 kN/mm

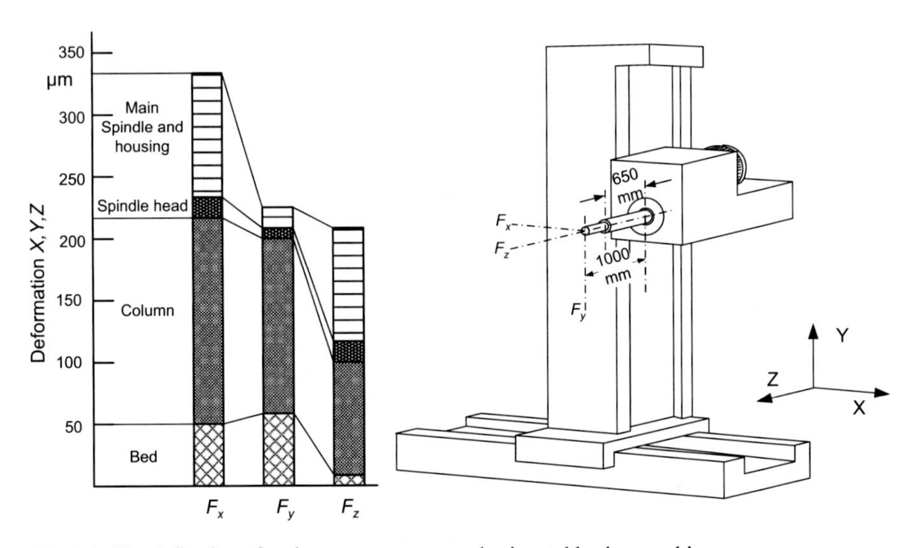

Fig. 5.2 The deflection of major components on a horizontal boring machine

Solution
Note the line represents the force flux, which identifies the parallel and serial elements.

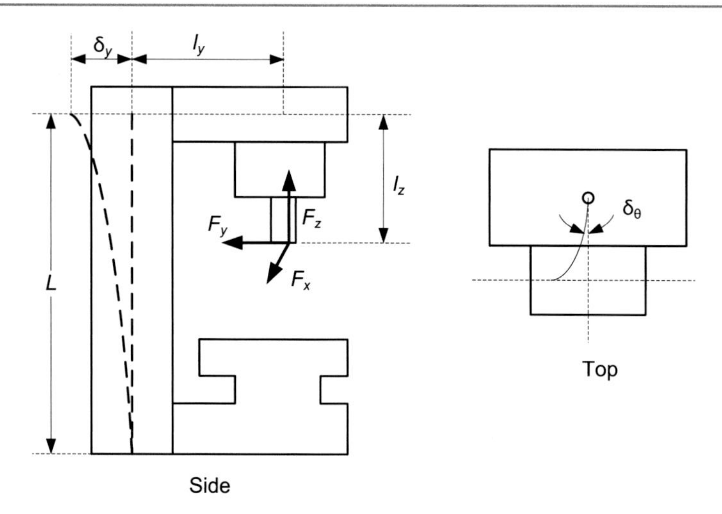

Fig. 5.3 Column static deflection of a vertical machine tool under 3-D cutting forces

$$k_{total} = \left[\frac{1}{k_1} + \frac{1}{k_2} + \frac{1}{k_3} + \frac{1}{k_4} + \frac{1}{2k_5} + \frac{1}{k_6} + \frac{1}{k_7 + k_8} + \frac{1}{k_9}\right]^{-1}$$

$$= \left[\frac{1}{45} + \frac{1}{20} + \frac{1}{62} + \frac{1}{25} + \frac{1}{2(78)} + \frac{1}{120} + \frac{1}{37 + 42} + \frac{1}{58}\right]^{-1} = 5.77 \text{ kN/mm}$$

The structural design of a machine tool needs to also factor in the bending and torque loads. Considering the columns as beans, elastic stress analysis can be applied to simple cross-section geometries to calculate the machine's resistance to bending and torsion. Figure 5.3 shows the column deflection of a vertical machining center as a three-dimensional (3-D) cutting force is produced at the tool–work contact.

The deflection in the Z direction involves the elongation of L due to F_z, the shortening of l_z due to F_z, the uplifting of l_y due to F_z, and the bending rotation of the spindle head overhang due to a moment caused by F_y:

$$\delta_z = F_z\left(\frac{L}{AE} + \frac{l_z}{A_s E_s} + \frac{l_y^3}{3E_h I_{h,z}}\right) - F_y l_z\left(\frac{l_y^2}{2E_h I_{h,z}}\right) \tag{5.1}$$

where A and E are the cross-sectional area and modulus of elasticity of the frame, A_s and E_s are those of the spindle, and E_h and $I_{h,z}$ are the modulus of elasticity and the area moment of inertia of the spindle head (overhang). Note that $I_{h,z} = \int_A z^2 dA$ is the area moment of inertia about the neutral axis, namely the X axis passing through the geometrical center of the X-Z cross section of the spindle head.

The deflection also has a translation component on the Y-Z plane, δ_y, associated with the displacement of the geometrical center on the X-Y cross section. The other translation component is on the X-Z plane, δ_x, associated with the same center, and an angular component is on the X-Y plane, δ_θ. They can be calculated from:

$$\delta_y = \frac{M_x}{\left(\frac{2EI_y}{L^2}\right)} + \frac{F_y}{\left(\frac{3EI_y}{L^3}\right)} + \frac{F_y l_y}{A_h E_h} + \frac{F_y}{\left(\frac{3E_s I_{s,y}}{l_z^3}\right)} \tag{5.2}$$

where $M_x = F_z l_y - F_y l_z$ and $I_y = \int_A y^2 dA$ which is the area moment of inertia about the neutral axis, namely the X axis passing through the geometrical center of the X-Y cross section of the frame. The terms in Eq. (5.2) are attributed to the bending of the column, the concentrated side loading on the column, the shortening of the overhang, and the concentrated side loading on the spindle assembly. The deflection in the X direction is

$$\delta_x = \frac{F_x}{\left(\frac{3EI_x}{L^3}\right)} - \frac{F_x l_z}{\left(\frac{2EI_x}{L^2}\right)} + \frac{F_x}{\left(\frac{3E_h I_{h,x}}{l_y^3}\right)} + \frac{(F_x l_y)L l_y}{GJ_z} + \frac{F_x}{\left(\frac{3E_s I_x}{l_z^3}\right)} + \frac{(F_x l_z) l_y l_z}{GJ_y} \tag{5.3}$$

where $I_x = \int_A x^2 dA$ which is about the Y axis passing through the geometrical center of the X-Y cross section of the frame, $J_z = \int_A (x^2 + y^2) dA$ is the polar area moment of inertia with respect to the Z axis passing through the geometrical center, and $J_y = \int_A (x^2 + z^2) dA$ is the polar area moment of inertia with respect to the Y axis passing through the geometrical center. The terms in Eq. (5.3) are results of the concentrated side loading on the column, bending on the column, concentrated side loading on the overhang, rotation of the column, concentrated side loading on the spindle assembly, and the rotation of the overhang. Note that the fourth term is depicted in Fig. 5.3 with $T_\theta = F_x l_y$

$$\delta_\theta = \frac{T_\theta L}{GJ_z} \tag{5.4}$$

Note that this assumes that components other than the column and the spindle are perfectly rigid. Analysis of other component deflection can be added to the column deflection based on elastic superposition. The above analysis also does not consider the effect of shear stresses, which is expected to be relatively small anyway. In addition, the analysis does not include the weight of the spindle, housing, and tool. These elements can contribute to the deflection of the column, but only in the sense

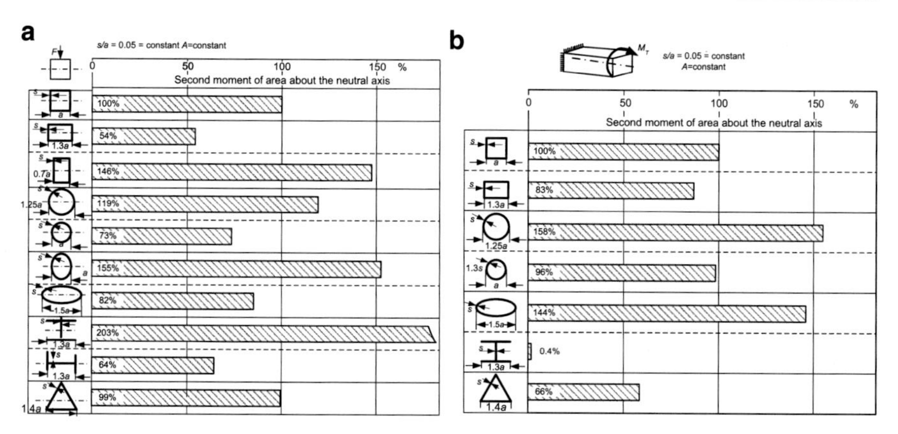

Fig. 5.4 Comparison of the **a** area moments of inertia, and **b** polar area moments (of inertia for machine tool frame design)

Fig. 5.5 Various cross-section designs for machine tool column

of deadweights. Once the tool setting is completed prior to cutting, the deadweights do not play a role in the deflection-induced part form error.

Since the resistance to bending of the machine column is proportional to its area moment of inertia about the neutral axis, the cross-sectional geometry of a column is an important machine design parameter. Figure 5.4a lists various area moments of inertia of different cross sections. On the other hand, various polar area moments of inertia are listed in Fig. 5.4b. It should be noted that a section showing strong resistance to bending deflection does not necessarily resist torsion deflection better. For example, the oval cross section (the 7th in (5.4a) and the 5th in (5.4b)) has a poor resistance to bending but high resistance to torsion, while the I-beam (the 8th in (5.4a) and the 6th in (5.4b)) shows a very good resistance to bending but nearly zero resistance to torsion. Generally speaking, the bending as well as the torsion resistance of machine frame components may be enhanced by the provision of strengthening ribs as shown in Fig. 5.5. The relative bending and torsion resistance of these ribbed frames are summarized in Fig. 5.6.

In light of the large amount of effort involved in the analysis of complicated machines, finite element simulations are often used as a design tool to estimate machine deflections.

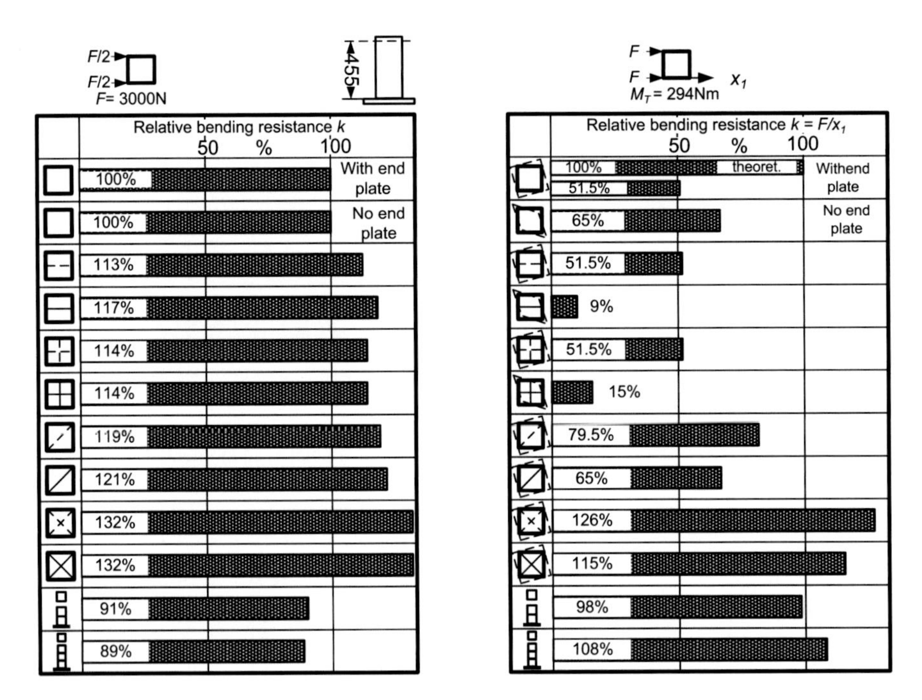

Fig. 5.6 Bending and torsion resistance for machine tool frames with various ribbing

Cutting forces often exhibit certain extent of time variation due to the engagement and disengagement of the cutter as discussed in previous chapters. The effect of machine tool dynamic characteristics on vibration behavior can be understood by considering a single degree of freedom, second-order dynamical system. The equation of motion of the system under a periodic excitation of magnitude F and frequency ω is $m\ddot{x} + c\dot{x} + kx = F_o \sin \omega t$. The particular solution to the above equation is a steady-state oscillation of the same frequency ω as that of the excitation:

$$X = \left(\frac{F}{k}\right) \frac{\sin\left\{\omega t - \tan^{-1}\left[\dfrac{c\omega / k}{1-\left(m\omega^2 / k\right)}\right]\right\}}{\sqrt{\left(1-\dfrac{m\omega^2}{k}\right)^2 + \left(\dfrac{c\omega}{k}\right)^2}} \tag{5.5}$$

The damping coefficient c plays a role in reducing the magnitude of forced vibration. When considering the machine tool frame, not just the damping properties of individual components but also the damping at the interfaces are of great importance. While the damping properties of steel are lower than those for cast iron, the damping effect at the welded joint of fabricated steel constructions generally com-

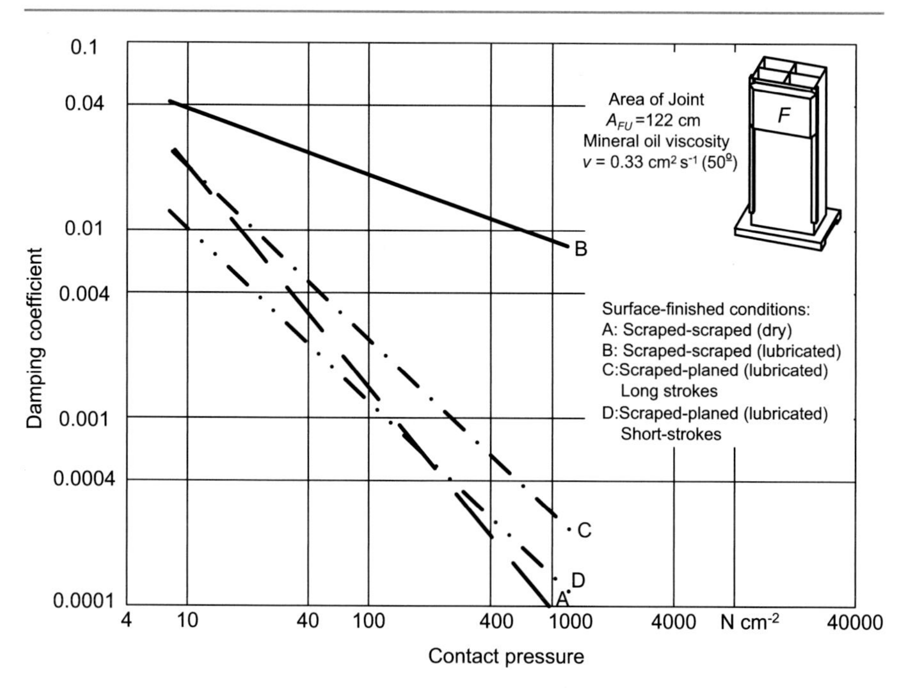

Fig. 5.7 The effect of surface finish and contact pressure on damping coefficient

pensates for this disadvantage. At a joint, the geometry of joint, the surface finish, the contact pressure, and media between the joint faces are all important factors.

Figure 5.7 shows the influence of surface finish and contact pressure at a column joint on the damping effect. The complicated mechanism of damping at a joint is not yet fully understood and no mathematical relationship for its dynamic property is available. Helpful steps to improve the damping properties of a system are, in general, to provide friction plates of lubricated scraped–scraped surfaces and to install auxiliary vibration absorbers at locations where large vibration amplitudes are expected.

5.2 Slides and Rails

Guideways are frame components that carry the workpiece table or spindles. There are two way types: plain or box and linear or rolling elements. Each consists of a slide riding on a track cast into, or bolted into the frame.

The slide carries the workpiece table or a spindle. The oldest and simplest is the box way. As a result of the large surface contact area, it has high stiffness, good damping characteristic, and high resistance to cutting and shock loads. The downside of the box slide is the stick-slip condition caused by the difference between dynamic and static friction coefficients that exists in box ways. This condition introduces positioning and feed motion errors. Hydrodynamic, hydrostatic,

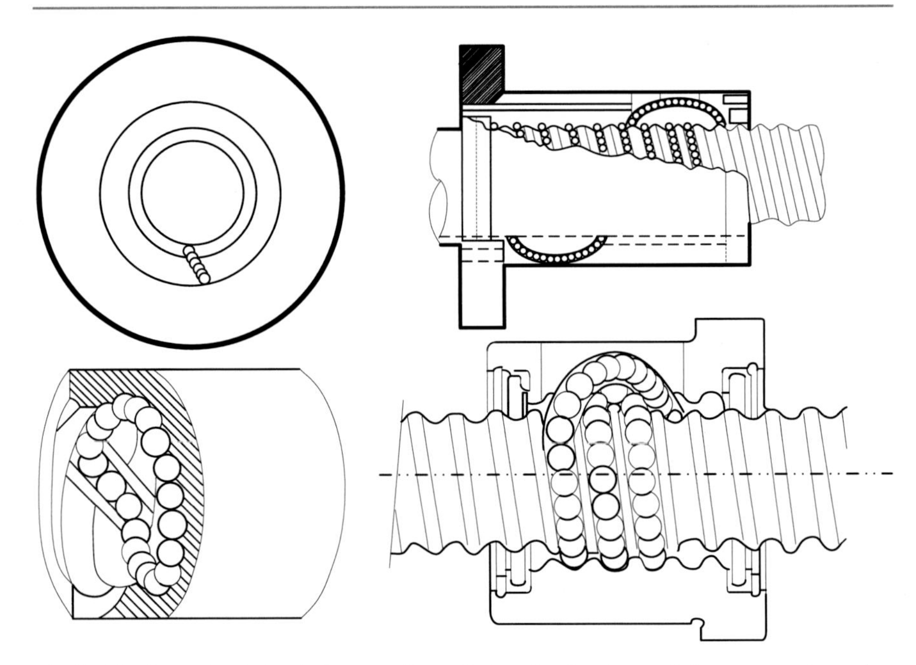

Fig. 5.8 In the recirculating ball screw, the balls travel a few turns of the screw, and are then picked up and deposited at the other end of the nut. This creates a continuous roller-bearing surface between the nut and the screw

aerodynamic, or aerostatic lubrications are often used to minimize this condition. The contacting surfaces on the box slide and rail have to be scraped to ensure a flat surface. Machine tool builders use box ways with a number of configurations including square, T, and V. Load direction and magnitude determine the best configuration for a given condition. For example, V shape is highly accurate but it does not carry large lateral load.

The linear way type also consists of a rail and a slide, but has a rolling-element bearing between the two. Rolling-element slides eliminate the stick-slip condition. Linear ways are lighter in weight and operate with less friction, so they can be positioned faster and with less energy. However, they are less robust due to the limited contact area. Each axis for a machine tool will typically have two rails bolted to the frame, and at least four bearing blocks to carry either a spindle or a table.

Slides are moved by hydraulics, rack-and-pinion systems, or leadscrews. Hydraulic pistons are the least costly, most powerful, most difficult to maintain, and the least accurate option. Heat buildup often reduces accuracy in these systems significantly. Motor-driven rack-and-pinion actuators are easy to maintain and are used where there is a large-motion range, but they can have accuracy problems and require a lot of power to operate. Motor-driven screws are the most common actuation method. The screws can either be leadscrews or ball screws, with the former being less expensive and the latter more accurate. A ball screw system is shown in Fig. 5.8. The recirculating ball screw has very tight backlash, thus it is ideal for

CNC machine tools since their tool trajectories are often piecewise continuous. A disadvantage of the ball screw systems is the stiffness due to limited contact area between the balls and the thread.

5.3 Motors

Electric motors are the prime movers for most machine tool functions. They are made in a variety of types to serve three general machine tool needs: spindle power, slide drives, and auxiliary power. Most of them use 3-phase AC power supplied at 220 or 460 V.

The design problem through the years with machine tools and motors has been how to get high torque at a variety of speed. Initially, mechanical transmissions consisting of gears, belts, and gear/belt combinations gave speed-changing capability. Up to 36 speed ranges were common at one time. But all this extra hardware is costly and needs maintenance. In recent years, the operational speed of spindles has risen significantly. For example, their conventional speed 10 years ago was around 1600 rpm. Today, they can turn at 15,000 rpm and higher. Higher speeds cause vibration, which makes the complex mechanical transmission unacceptable. Thanks to the dramatic improvement in motor design and control technology, it is now possible to quickly modify motor speed and torque. Mechanical systems involving more than three-speed transmission are becoming unnecessary for most high-speed and low-torque machines.

Motors for spindles are rated by horsepower, which generally ranges from 5 to 150 HP (3.7–112 kW) with the average around 50 HP (37 kW). Positioning motors, or feed drive motors, are usually designated by torque, which generally ranges from 0.5 to 85 lb-ft (0.2–115 Nm). The design of the main spindle motor requirement is based on the calculation of cutting force and power, as functions of material properties and cutting parameters, as discussed in Chap. 2. However, the design of feed drive motors has to consider both the static and dynamic loads associated with the cutting forces and the friction forces on the drive mechanisms.

When the machining table is moving at a constant speed, no acceleration and thus no inertia force exists and the feed drive load is static. For feed drive motors, the sources of the static load are basically the friction resistance in the guideways and the bearings plus possible cutting forces acting in the feed direction of the machining table (in the case of a milling machine for example). In addition, the feed drive motor has to provide enough torque to accelerate the table, workpiece, and leadscrew mechanism during a short time period until the table reaches the specified speed in a state of dynamic acceleration. The calculation of these static and dynamic torque requirements is discussed below.

There are three sources of static loads on machine tool feed drive: the friction in the guideways, friction in the feed drive bearings (thrust bearings and the nut), and cutting forces applied on the leadscrew shaft. Figure 5.9 shows the locations of these sources.

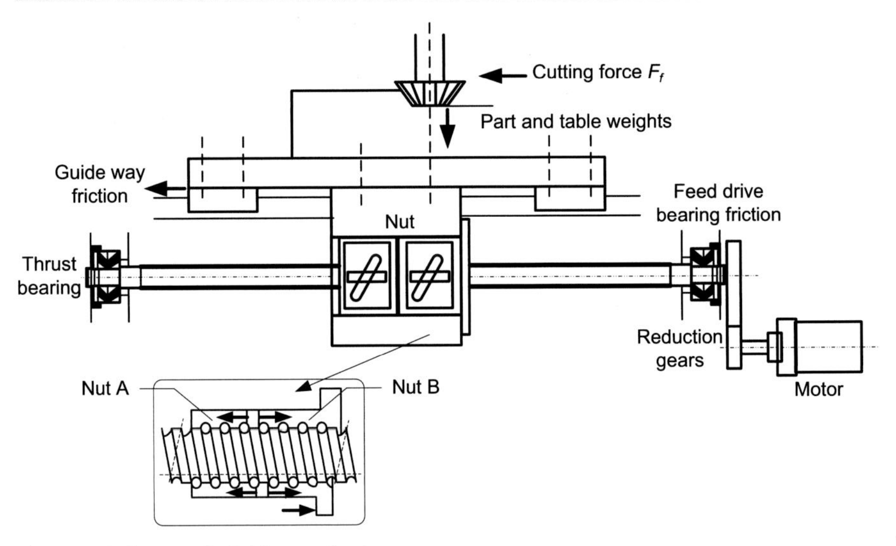

Fig. 5.9 Ball screw feed drive mechanism

The friction in guideways depends on the nature of contact between the slide and the rail. The torque consumed to overcome this friction is T_1.

$$T_1 = \frac{l}{2\pi}\mu_g\left(w_t + w_w + F_z\right) \tag{5.6}$$

in which l is the feed screw pitch, μ_g is the friction coefficient on the guideway, and it typically ranges from 0.05 (roller bearings) to 0.1 (plain lubricated slides). The w_t and w_w are the weights of the machining table and the workpiece respectively, and F_z is the cutting force perpendicular to the machining table.

Axial thrust bearings are used at both ends of the leadscrew to resist the feed forces. They are preloaded in tension to offset backlash in back and forth motions. Therefore, the torque required to overcome the bearing friction is

$$T_2 = \mu_b \frac{d_b}{2}(F_f + F_p) \tag{5.7}$$

where μ_b is the friction coefficient on the thrust bearings (0.002~0.005)—which include both the thrust bearings at the shaft ends and the nut friction on the shaft, d_b is the mean bearing diameter, F_f is the feed force, which can be estimated from cutting mechanics, and F_p is the preload.

The torque required to overcome the cutting force applied on the leadscrew shaft is

$$T_3 = \frac{l}{2\pi}F_f \tag{5.8}$$

The total static torque required is thus the sum of these individual torque components:

$$T_s = T_1 + T_2 + T_3 \tag{5.9}$$

If the required torque is too large, a reduction gear mechanism can be used at the cost of slide movement speed. A reduction gear, typically applied between the motor shaft and the leadscrew, is characterized by its reduction r of

$$r = \frac{N_s}{N_m} = \frac{\omega_m}{\omega_s} \tag{5.10}$$

where N_s is the number of teeth on the feed screw gear and N_m is that on the motor's gear, ω_m is the motor's gear rotational speed and ω_s is the feed screw gear rotational speed. When a reduction ratio greater than one is used, the required torque on the motor's shaft is reduced to

$$T'_s = \frac{T_s}{r} \tag{5.11}$$

Since the static torque refers to the torque in steady state operation, the feed drive motor needs to have a continuous, but not a peak, torque capability greater than T'_s.

The consideration of dynamic loads for feed drive motor is a result of acceleration torque when the table speed changes. The dynamic loads therefore consist of the inertia effects of the table, the workpiece, the leadscrew, the gears, and the motor's shaft. The moment of inertia of the table and workpiece with respect to the leadscrew shaft is

$$J_1 = (m_t + m_w)\left(\frac{l}{2\pi}\right)^2 \tag{5.12}$$

The moment of inertia of the leadscrew with a pitch diameter of d_p is

$$J_2 = \frac{1}{2} m_l \left(\frac{d_p}{2}\right)^2 \tag{5.13}$$

in which m_l is the mass of the leadscrew shaft. The total inertia with respect to the motor's shaft is then

$$J = \frac{J_1 + J_2}{r^2} + J_3 \tag{5.14}$$

where J_3 is the inertia of the motor's shaft itself and r is the reduction ratio between the feed screw and motor gears.

Another friction torque that needs to be considered is the viscous friction torque proportional to the velocity, particularly in the reduction gear box with lubricant. The total dynamic torque required to accelerate the inertia J and to overcome viscous friction and the static loads is given as

$$T = J\frac{d\omega}{dt} + B\omega + T'_s \tag{5.15}$$

where ω is the motor angular velocity and B is the viscous friction coefficient. The peak torque delivered by the motor must be larger than what is calculated from above, although a gear reduction can be applied to lower the torque requirement at the expense of achievable feed speed.

Example

In selecting the longitudinal feed drive servomotor capability (torque requirement) for a vertical machining center, the following variables are considered: max workpiece mass = 200 kg, machining table mass = 180 kg, leadscrew mass = 8.15 kg, feed screw pitch = 0.00508 m/rev, feed screw diameter = 0.0445 m, motor shaft inertia = 2.373×10^{-3} kg-m^2, no gear reduction, guide friction coefficient = 0.1, bearing friction coefficient = 0.005, max vertical cutting force = 2000 N, max feed cutting force = 8000 N, preload force in thrust bearings = 5000 N, rapid traverse velocity = 0.17 m/s, servo rise time = 0.1 s, viscous damping coefficient = 0.015 Nm/(rad/s), desirable safety factor = 2 under both static and dynamic conditions.

Solution

$$T_1 = \frac{l}{2\pi}\mu_g(w_t + w_w + F_z) = \frac{0.00508}{2\pi}(0.1)[(180+200)9.8+2000] = 0.463\ \text{Nm}$$

$$T_2 = \mu_b\frac{d_b}{2}(F_f + F_p) = 0.005\frac{0.0445}{2}(5000+8000) = 1.446\ \text{Nm}$$

Note that the mean bearing diameter is taken here as the feed screw diameter, although this will not be the case if a larger thrust bearing diameter is used.

$$T_3 = \frac{l}{2\pi}F_f = \frac{0.00508}{2\pi}8000 = 6.468\ \text{Nm}$$

Total required continuous torque is
$T_s = T_1 + T_2 + T_3 = (0.463 + 1.446 + 6.468] \times 2 = 16.78$ Nm

$$J_1 = (m_t + m_w)\left(\frac{l}{2\pi}\right)^2 = (180+200)\left(\frac{0.00508}{2\pi}\right)^2 = 2.484\times10^{-4}\ \text{kgm}^2$$

$$J_2 = \frac{1}{2}m_l\left(\frac{d_p}{2}\right)^2 = \frac{1}{2}(8.15)\left(\frac{0.0445}{2}\right)^2 = 20.174\times10^{-4}\ \text{kgm}^2$$

$$J = \frac{J_1 + J_2}{r^2} + J_3 = 2.484\times10^{-4} + 20.174\times10^{-4} + 2.373\times10^{-3} = 4.64\times10^{-3}\ \text{kgm}^2$$

The linear acceleration of the table can be calculated by dividing the rapid feed velocity by the servo's expected rise time:

$$a = \frac{0.17}{0.1} = 1.7\ \text{m/s}^2$$

The angular acceleration of the motor's shaft is

$$\frac{d\omega}{dt} = \frac{a}{(l/2\pi)} = \frac{1.7}{(0.00508/2\pi)} = 2{,}103\ \text{rad/s}^2$$

The total torque requirement is found by considering the maximum angular speed:

$$\begin{aligned} T &= J\frac{d\omega}{dt} + B\omega + (T_1 + T_2 + T_3) = J\frac{d\omega}{dt} + B\left(\frac{0.17}{l/2\pi}\right) + (T_1 + T_2 + T_3) \\ &= \{4.64\times10^{-3}\ \text{kgm}^2 \times 2103\ \text{rad/s}^2 \\ &+ [0.015\ \text{Nm/(rad/s)} \times 0.17(2\pi/0.00508)\ \text{rad/s}] \\ &+ (0.4631 + 1.4463 + 6.468)\}\text{Nm} \times 2 \\ &= 42.58\ \text{Nm} \end{aligned}$$

Therefore, the feed drive motor has to be able to deliver a 42.58 Nm torque for a period of at least 0.1 s. Note that the dynamic torque required is about 250% of the static requirement, depending on the system inertia and demanded acceleration. However, the use of maximum acceleration to reach a maximum speed typically takes place only in a non-cutting condition where

$$T_1 = \frac{l}{2\pi}\mu_g(w_t + w_w) = \frac{0.00508}{2\pi}(0.1)\,[(180+200)9.8] = 0.301\ \text{Nm},$$

$$T_2 = \mu_b\frac{d_b}{2}F_p = 0.005\frac{0.0445}{2}(5000) = 0.556\ \text{Nm},$$

and T_3=0.

$$
\begin{aligned}
T &= J\frac{d\omega}{dt} + B\omega + (T_1 + T_2) = J\frac{d\omega}{dt} + B\left(\frac{0.17}{l/2\pi}\right) + (T_1 + T_2) \\
&= \{4.64\times10^{-3}\ \text{kgm}^2 \times 2103\ \text{rad/s}^2 \\
&+ [0.015\ \text{Nm/(rad/s)} \times 0.17(2\pi/0.00508)\ \text{rad/s}] \\
&+ (0.301 + 0.556)\}\ \text{Nm} \times 2 \\
&= 27.54\ \text{Nm}
\end{aligned}
$$

which is 164% of the static torque requirement.

5.4 Spindles

The spindle is responsible for delivering torque to the cutting tool. Its functions are (1) to guide the tool and/or workpiece at the cutting point with adequate kinematic accuracy, and (2) to absorb externally applied forces such as the workpiece weight and cutting forces with minimum static, dynamic, and thermal distortions. In all, the precision of a spindle is essential to the overall machine tool operation. The key factors influencing precision are bearing type and placement, lubrication, and cooling.

The least complex spindle bearing design is the hydrodynamic, which is a journal bearing. The shaft turning action draws oil between the metal surfaces. Low friction is achieved this way at sufficiently high speeds. However, at starting or in low-speed operations the metal surfaces may not be completely separated due to the lack of dynamic driving forces. Hydrostatic bearing comes into play in this situation by introducing the lubricant into the load-bearing area at a pressure high enough to separate the metal surfaces with a relatively thick film of lubricant. So, unlike hydrodynamic lubrication, it does not require motion of one metal surface relative to another. The downside of hydrostatic bearing is the auxiliary external fluid supply system that is bulky, maintenance demanding, and expensive.

Air is sometimes used as a lubricant in either aerostatic or aerodynamic bearings. Being noncontact bearings, they work well in precision, low-load applications. Another noncontact design is the magnetic bearing in which carefully balanced magnetic fields keep the shaft centered in its housing. These bearings are virtually frictionless so there is no shaft wear. The main disadvantages are initial cost, large size, and potential maintenance difficulty.

The use of rolling-element bearings is also rather common in machine tool spindles. In this category, there are tapered roller bearings and ball bearings. The former has better stiffness (over 1000 N/μm), which is ideal for high-axial loads, but more friction than the latter. The highest speed spindles often use hybrid ceramic ball bearings combining steel races with ceramic balls that are about 60% lighter than steel, and so go faster. One of the considerations in high-speed applications is the thermal breakdown and seizing, and ceramic balls have low heat retention rate and low thermal expansion to resist these difficulties.

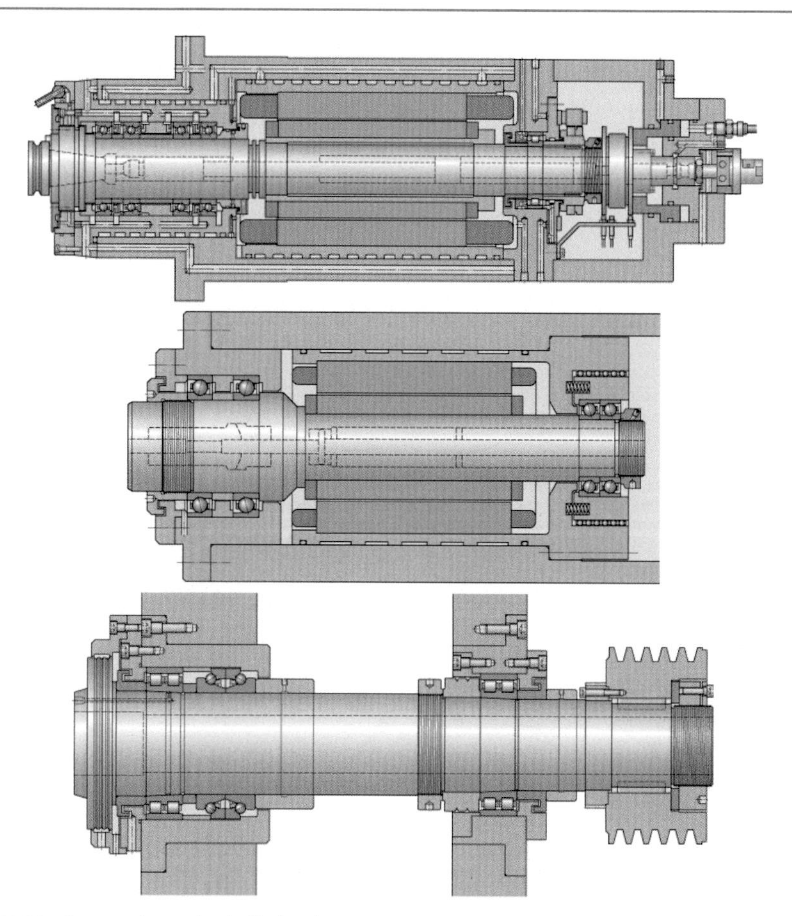

Fig. 5.10 Commonly used spindle bearing designs. (Source: SKF)

Figure 5.10 shows a number of typical spindle-bearing configurations. The stiffness of the bearings in the figure reduces from the upper to the lower diagram, while acceptable speed range increases. The stiffness of bearing affects the machined part form accuracy in both the radial and axial directions. The axial stiffness of a spindle, especially with the use of preloaded taper roller type, is considerably greater than that of the bearing, so attention should be directed to the stiffness in the radial direction.

In terms of radial stiffness, the total deflection of a spindle system at the point of contact between tool and workpiece can be regarded as the result of an overhanging load with spring support, one at the collar end and the other at the drive end. Figure 5.11 illustrates the principle of elastic superposition for the estimation of cutter point of contact deflection attributed to the spindle flexure and the bearing flexure. The former is

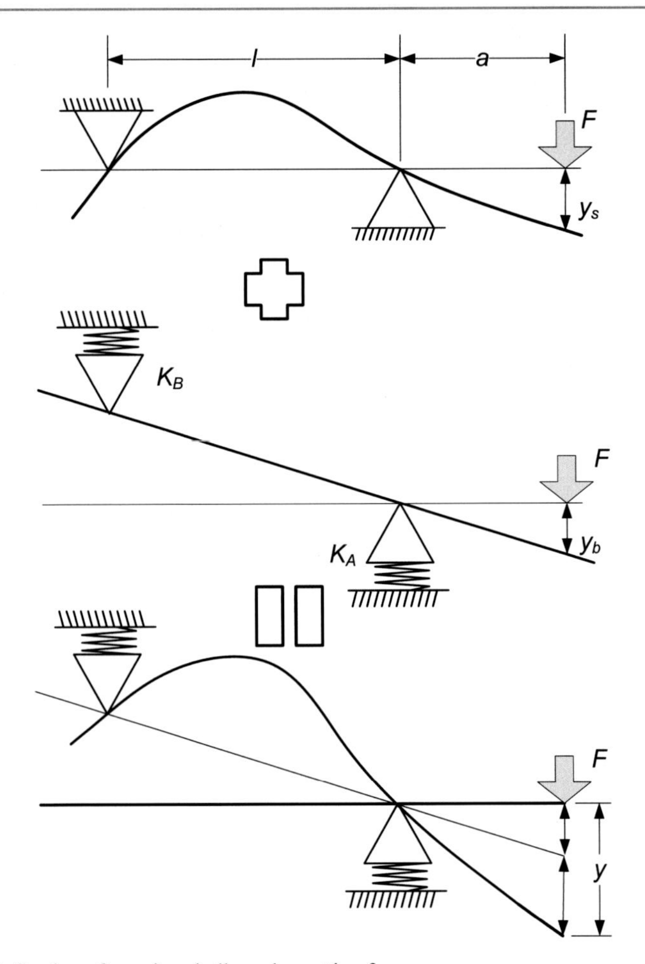

Fig. 5.11 Deflection of a main spindle under cutting force

$$y_s = \frac{F}{k_s}, \text{ where } k_s = \frac{3EI}{a^2(l+a)} \tag{5.16}$$

and the latter is

$$y_s = \frac{F}{k_b}, \text{ where } k_b = \frac{l^2}{\dfrac{(a+l)^2}{K_A} + \dfrac{a^2}{K_B}} \tag{5.17}$$

Note that E is assumed constant over the length of the spindle.

The total flexure is then

$$y = y_s + y_b = \frac{F}{k}, \text{ where } k = \frac{k_s k_b}{k_s + k_b} \tag{5.18}$$

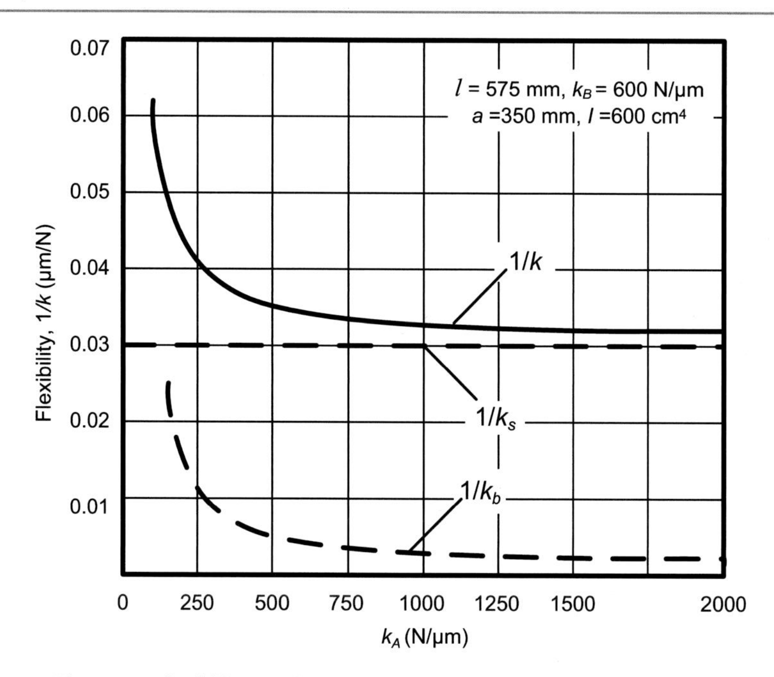

Fig. 5.12 The system flexibility as a function of collar-end bearing stiffness

Figure 5.12 shows the contribution of the collar-end bearing stiffness K_A to the total flexibility of a particular spindle. For this case, the overall flexibility does not change with the bearing if K_A is greater than 750 N/μm.

Compared with hydrodynamic bearings, rolling-element bearings, in general, have the major advantage of low starting friction (typically coefficient of friction between 0.001 and 0.002). Ball and roller bearings can be preloaded by pressing together mating bearing elements rather than operating with a clearance. This implies a better precision of the spindle shaft. On the other hand, hydrodynamic bearings are well suited for high rotating speeds with impact and momentary overloads. The higher the rotating speed the more effective the hydrodynamic lifting action. Also, the fluid film effectively "cushions" lateral impact, as the duration involved is too short for the film to be squeezed out. High rotating speeds are generally disadvantageous to rolling-element bearings because of the rapid accumulation of fatigue cycles and the high centrifugal force on the rolling element. Rolling-element bearings also occupy more space radially. They are generally noisier as well.

The supply of lubricant is important since bearings often seize immediately after the failure of lubrication system. In addition, lubricant controls heat buildup in the spindle bearing which, in turn, influences accuracy. The most widely used for lower speed application is grease pack. It involves packing the bearing with grease that is held in with a seal. It is simple and often maintenance free for extended periods. As speeds go up, however, temperature rises and the grease melts. Commonly used methods for lubricating higher speed bearings include oil mist and oil jet. These methods flood the bearing race with oil under pressure, and they perform cooling function as well. Speed up to 200,000 rpm is permissible; however, it can cause environmental problems if the oil blows through the machine into the plant.

At high speed, heat buildups in a spindle can come from the bearings, the motor, or even the tool. Although lubrication removes a portion of the heat, high-power spindles often have a water jacket around the outer bearing races as a separate cooling system. The water jacket can have temperature control, which supplies heated water to bring the spindle to its optimum operation temperature quickly and cooling water to increase motor life and efficiency.

5.5 Tool Materials

The selection of cutting tool materials is one of the most vital factors for the effectiveness of the process. During cutting, the tool usually experiences high temperatures, high stresses, rubbing friction, sudden impact, and vibrations. Therefore, the two important factors in the selection of cutting tool materials are hardness and toughness. Hardness is the endurance to plastic deformation and wear, and it is especially important at elevated temperatures. Toughness is the resistance to impact and vibrations, which are rather common in interrupted cutting such as milling. However, hardness and toughness do not go together; thus, the selection of cutting tool is often a matter of compromise.

Cutting tool materials have gone through generations of evolution. Carbon steels of 0.9–1.3 % carbon and alloying elements such as Mo and Cr were called tool steels and were once the most common cutting tool materials. They lose hardness at temperatures above 400 °F because of tempering and have largely been replaced by high-speed steel (HSS) since the early 1900s. HSS typically contains 18 % W or 8 % Mo, and other elements such as Co and Cr. It retains its hardness at temperatures up to 1100 °F and can operate at about double the cutting speed with equal life. The term "high-speed" steel has historical reason. In today's technology, high-speed machining refers to the process showing either a cutting temperature above 1000 °C, a cutting speed higher than 100 m/s, or a spindle speed higher than 50,000 rpm, although this definition is somewhat dependent upon the workpiece material involved as shown in Fig. 5.13. Apparently, none of these conditions can be met with the use of HSS cutters anymore. Both tool steels and HSS are tough and resistive to fracture; therefore, they are ideal for processes involving interrupted engagements and machine tools with low stiffness that are subject to vibration and chatter.

Cast cobalt alloys, popularly known as Stellite tools, are cobalt-rich, chromium–tungsten–carbide cast alloys. Though comparable in room temperature hardness to HSS tools, cast cobalt alloy tools retain their hardness to a much higher temperature, and they can be used at 25 % higher cutting speeds than HSS tools. They are available only in simple shapes, such as single-point tools and saw blades, and are being phased out because of increasing cost due to the shortages of strategic raw materials (Co, W, and Cr).

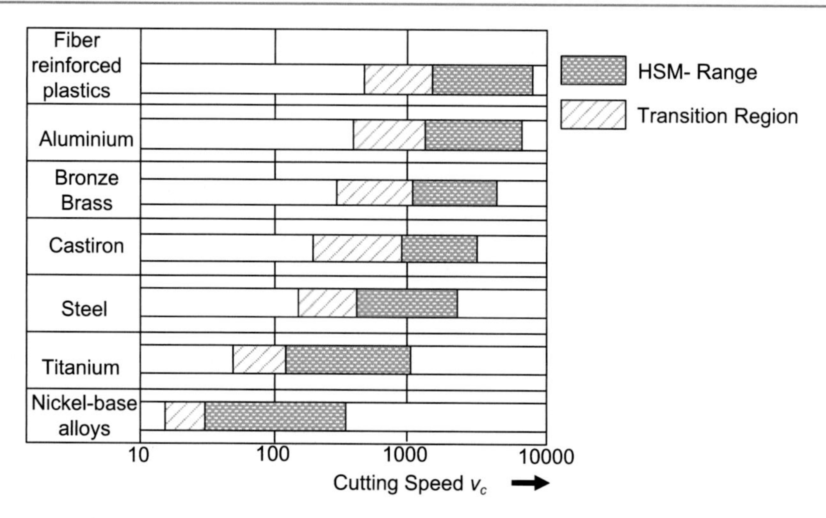

Fig. 5.13 The definition of high-speed machining based on cutting speed and workpiece materials. (Source: Schulz et al., High-speed machining, CIRP Annals, 41(2): 637–645, 1992)

Carbides became popular during World War II as they offered a four- or five-fold increase in cutting speeds over HSS. Most carbide tools in use are WC based, although some TiC-based tool materials have also been developed, primarily for auto industry applications. They are composite materials, consisting of WC or TiC particles bonded together in a cobalt matrix; hence, they are referred to as cemented carbides. These carbide tools are much harder and chemically stable, however, more brittle and less tough. They are good for continuous roughing on rigid machines, but should avoid shallow cuts, interrupted cuts, and compliant machines because of likely chipping. In very low-speed cutting, the chips tend to weld to the tool face and cause microchipping of the edge. Carbide tools are expensive but are available in insert form. When one edge becomes dull, the insert can be rotated or turned over for a new edge.

About 5–10 μm of TiC, TiN, or Al_2O_3 can be coated over HSS or carbide substrate by chemical or physical vapor deposition. The coatings provide further resistance to abrasion, temperature, friction, and chemical reactions. These coated tools were introduced in the early 1970s, and have gained wide acceptance since then. Coated tools have two or three times the wear resistance of the best uncoated tools, thus providing a 50–100 % increase in speed for the same tool life. Currently, more than 50 % of the carbide tools used in the USA are coated.

Ceramic tools are primarily made of pure aluminum oxide, Al_2O_3, although black ceramics (cermets, 70 % Al_2O_3 and 30 % TiC) are also used. These tools can be operated at two to three times the cutting speeds of tungsten carbide, almost completely resist cratering, usually require no coolant, and have about the same tool life at their higher speeds as tungsten carbide does at lower speeds. However, ceramics lack toughness and therefore require more rigid tool holders and machines in order to take advantage of their capabilities. Interrupted cuts and interrupted application of coolants can lead to premature tool failure due to poor mechanical and thermal shock resistance.

Cubic boron nitride (CBN) is the hardest material presently available next to diamond. It is a synthetic material developed by General Electric by powder sintering. Its cost is somewhat higher than either carbide or ceramic tools but can cut five times as fast as carbide, and can hold hardness up to 1100 °C. It is ideal for machining hard aerospace materials such as Inconel 718, Rene 95, GTD-110, and chilled cast iron.

Single-crystal diamonds, with a cutting edge radius of 100 Å or less, are being used for precision machining of large mirrored surfaces. Industrial diamonds are now available in the form of polycrystalline compacts, which are finding industrial application in the machining of aluminum, bronze, and plastics, greatly reducing the cutting forces as compared to carbides. In addition to high hardness, unique features of diamond tools include good thermal conductivity, small cutting-edge radius, and low friction. Diamond machining is done at high speeds with fine feeds for finishing and produces excellent finishes. Shortcomings with diamond tools are brittleness, cost, and the tendency to interact chemically with workpiece materials such as carbon steel, Ti, and Ni.

Example

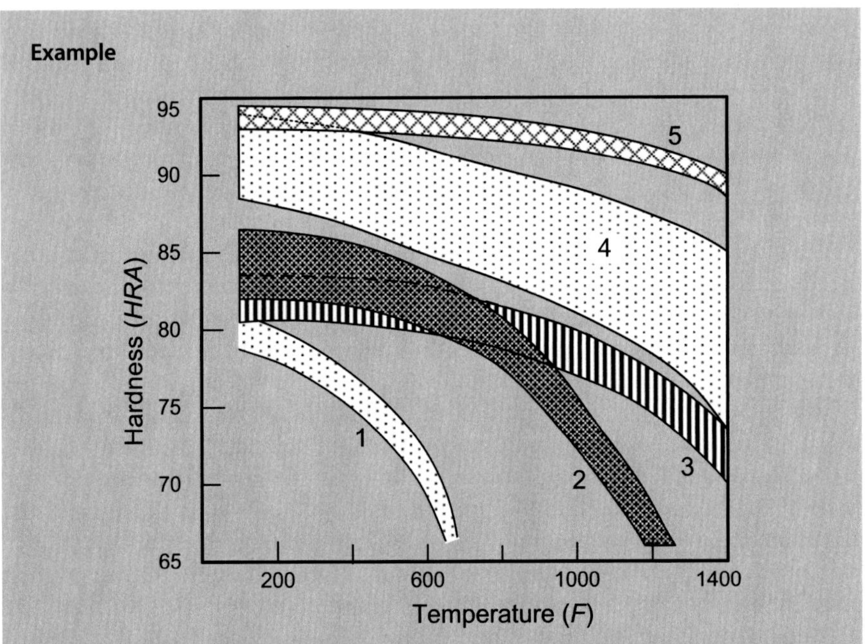

In the above figure the hardness of different tool materials is plotted as a function of temperature. Identify the tool materials for curves 1, 2, 3, 4, and 5.

If a block of cast iron workpiece is to be face milled with a three-flute cutter at 0.05 in. axial depth of cut, 0.01 in./tooth feed, and 500 ft/min surface speed, would you recommend the use of ceramic tools for this application? How about HSS tools? Why?

Solution
1-Carbon steel, 2-HSS, 3-Cast cobalt, 4-Carbide, and 5-Ceramic.

For the face milling, you do not want to use ceramic tool because of its lack of toughness in interrupted cutting. No HSS either since the temperature may go way higher than 1100 °F in this case.

Example
In the case of face milling carbon steel workpiece at 40,000 rpm with 5 in. diameter, 6 insert, cutter at 2 in. radial depth of cut and 0.05 in. axial depth of cut. What is your recommendation for tool material? Why?

Solution
Carbide insert tool. HSS would be too soft at the high speed, and ceramic would be too brittle for interrupted cutting.

5.6 Controls

Control components for machine tools have changed dramatically from limit switches to electric relays, to numerical control (NC), to programmable logic control (PLC), then to CNC. As computers got smaller and their power grew, they often reside in the machine as a dedicated and online control unit. Large-machining operations involving two or more machines are reverting, however, to a separate mainframe computer that is powerful enough to control several machine tools or an entire shop. The system is often referred to as distributed numerical control (DNC).

In a CNC system, the engineering prints are translated into machine-operating commands through programming. A programmer breaks the machining job into a series of data blocks. Each data block requires a certain number of commands and is specified by an alphabet code. The coding system has become an industry standard and is the basis for all commands today. For example, the S-codes give spindle speed, T-codes identify tools to be used, and G-codes are preparatory commands. As an alternative, the operating commands can be generated by input digitizing. A probe, laser light, or a mouse traces the surface of the part to be copied or the shape to be formed from an engineering drawing. The digitized information can be inverted, expanded, reduced in size, or blended with another part. In addition, computer-aided design programs can create or modify the geometrical model of parts. These inputs then go to computer-aided manufacturing (CAM) programs which factor in the allowance for tool size, interference with obstacles, and the elimination of unnecessary details to yield the tool paths. A lot of the CAM programs utilize expert systems. For example, the operator may know the workpiece material and the tool to use, but not the best speed. The program can suggest a speed based on other operating conditions.

Machine tools initially did simple single-axis moves involving one drive motor at a time. This was suitable chiefly for drilling or punching operations. State-of-the-

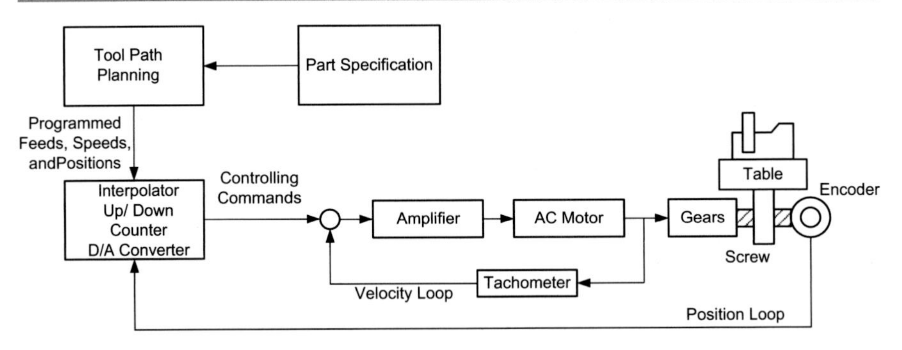

Fig. 5.14 The position and velocity feedback loops in a milling machine

art machines have control and actuating capability to perform multi-axis motion simultaneously. Machine tools with as many as 16 axes are available, with 3–5 axes being the norm. The simultaneous motion interpolates a continuous curvilinear path by small, straight-line increments. These increments are as small as 0.1 μm so that any kind of shape can be achieved.

Computer control depends on feedback for performance as shown in Fig. 5.14. Resolvers, encoders, and scales are often installed to measure the position and velocity of the workpiece or the tool. These sensors are expected to provide resolution at least one order of magnitude better than the machine actuating drives can respond to. For example, an encoder generating a million pulses per revolution produces typically 0.1 μm resolution, which is far more information than a position drive can use. Sensors also check for the presence or absence of a tool and, in some cases, its wear and breakage. Other sensors monitor the machine's vital signs, such as coolant flow, temperature, and whether auxiliary equipment (fans and pumps) is functioning.

Machine tool controls are developing in two new areas: communication and adaptive control. Expanded communication utilizes the data collected by machine control to tune other aspects of the manufacturing operation. For example, data on production time and number of parts produced can go into a data base for use in inventory control and quality monitoring. Or the same data can be adapted for production scheduling, part inspection, and error detection. Adaptive control enables the machine tool to modify process parameters leading to a higher productivity. For instance, tools are traditionally discarded based on average time in use, not actual performance. With adaptive control the machine senses when a tool is dull and will request a tool change. Traditional CNC programs are written conservatively to avoid any possible chatter vibrations. With adaptive control the machine detects the onset of chatter automatically and changes process parameters to eliminate that.

The term "adaptive control" commonly used in the machine tool industry is somewhat different than the term as used in the control community in the sense that it does not necessarily involve the variation of controller parameter during the process. The target of control in an adaptive control system is not limited to the feed rate and tool path as what traditional CNC systems do, but it often includes the power, force, temperature, vibration, and noise level as the final target since they relate to the functionality of the machining process. Two examples of adaptive control are to be discussed below. They involve the force control of a turning process and the power control of a cylindrical grinding process.

In the case of tuning, force is often an object to be maintained through feedback control. Force control achieves constant material removal rate and constant chip load such that the cutter, spindle, and the machine tool system as a whole, can be protected from overloading while maximum productivity is sought. Figure 5.14 shows a block diagram of a turning system under computer feedback control using feed rate as a manipulable variable. Note that to facilitate effective control the dynamics of the process has to be well understood. The dynamics of the process is defined by the relationship between the input (u_c) and the output (f_m) as

$$\frac{F_m(z)}{U_c(z)} = \left(\frac{U_a(z)}{U_c(z)}\right)\left(\frac{H(z)}{U_a(z)}\right)\left(\frac{F(z)}{H(z)}\right)\left(\frac{F_m(z)}{F(z)}\right) \tag{5.19}$$

where $F_m(z), U_c(z), U_a(z), H(z)$, and $F(z)$ are the z-transform of $f_m(t), u_c(t), u_a(z), h(z)$, and $f(t)$ as defined in Fig. 5.14. Note that the z-transform is always preferred in the computer control analysis and design since the computer operates in the discrete time domain. The four terms involved in the above equation are independent and have to be evaluated separately. The relationship between u_c and u_a depends on the equivalent mass, stiffness, and damping coefficient of the machine feed drive system. This relationship can often be experimentally identified by observing the feed rate command and measured actual feed rate simultaneously. For example, in a case reported by Zhang and Tomizuka (1988), a Tree UP-1000 lathe was used for turning control and the feed drive dynamics was observed to match well with a second-order transfer function:

$$\frac{U_a(z)}{U_c(z)} = \frac{0.2055z + 0.4529}{z^2 + 0.215z + 0.2466} \tag{5.20}$$

The relationship between the actual feed rate command and the chip thickness can be simply depicted by kinematics (Fig. 5.15). Figure 5.16 shows the evolution of chip thickness as a constant feed rate is applied.

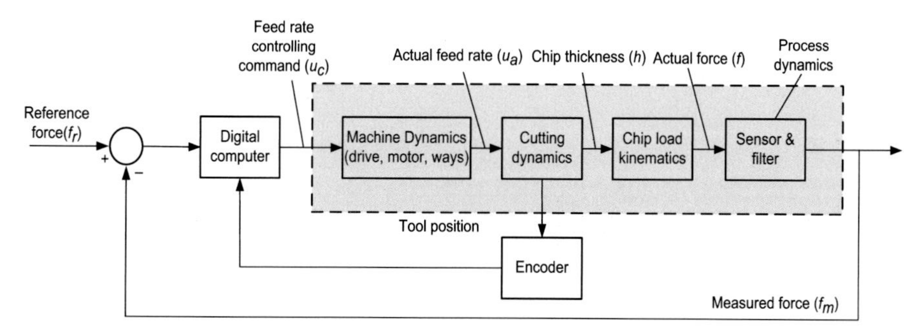

Fig. 5.15 Closed-loop control system of a machine

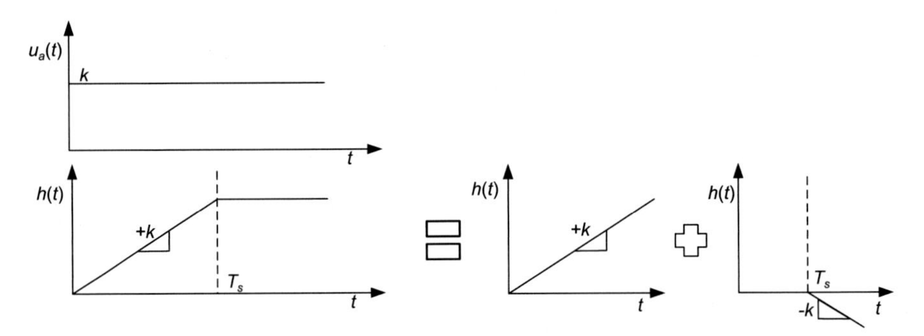

Fig. 5.16 Evolution of chip thickness

Therefore, $u_a(t)=\begin{cases}0, & t<0\\ k, & t\geq 0\end{cases}$, and $h(t)=kT-kt(t-T_s)$. From the basic principle of z-transformation, $U_a(z)=\frac{kz}{z-1}$ and $H(z)=kT\frac{z}{(z-1)^2}(1-z^{-T_s/T})$ where T is the sampling period of the computer controller,

$$\frac{H(z)}{U_a(z)}=\frac{T}{z-1}\left(1-z^{-\frac{T_s}{T}}\right) \tag{5.21}$$

From the discussion in Chap. 2, the relationship between the force and the chip thickness is well understood to be $f=p_s bh$, where p_s is the specific cutting energy and b is the radial depth of cut. So, $F(z)/H(z)=p_s b$.

The measurement of cutting force is typically done with the use of piezoelectric-based dynamometer, which has frequency response well above thousands of Hertz. Therefore, the dynamics of the dynamometer can be ignored. For a 10 ms sampling time, the pulse transfer function of the filter (known to have a dynamics of 1/(0.017s + 1)) is:

$$\frac{F_m(z)}{F(z)}=\frac{0.454}{z-0.5464} \tag{5.22}$$

In summary, the process dynamics is

$$\frac{F_m(z)}{U_c(z)}=0.454bp_sT\frac{\left(1-z^{-\frac{T_s}{T}}\right)(0.2055z+0.1783)}{(z-1)(z^2+0.215z+0.2466)(z-0.5464)} \tag{5.23}$$

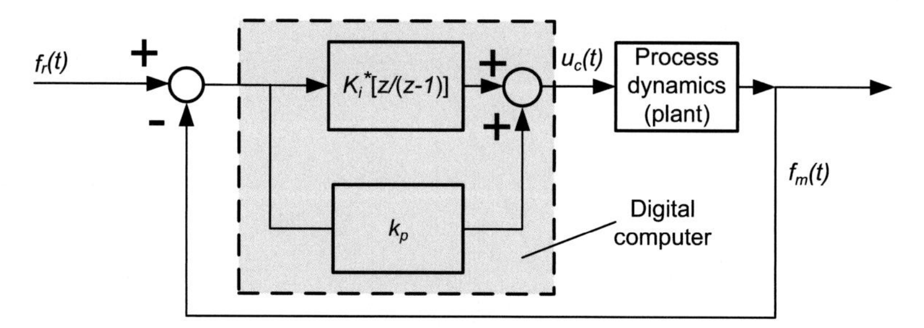

Fig. 5.17 Proportional and integral (PI) control of cutting force

This process dynamics model is useful for the design of feedback controllersas a standard proportional and integral (PI) controller of the following dynamic as shown in Fig. 5.17.

$$G_c(z) = k_p + k_i \frac{z}{z-1} \tag{5.24}$$

with $k_p = 1$ and $k_i = 0.67$, the closed-loop step response is as shown in Fig. 5.18.

In the case of cylindrical grinding, the benefits of exercising power or force feedback control can be appreciated from the viewpoint of accuracy and productivity. Figure 5.19 shows the time delay and positioning error of the infeed motion of a cylindrical grinding wheel in the case of feedforward position control. With the use of power/force feedback control the position command can be altered to accommodate for the dynamics of the grinding and the machine such that the target final position can be achieved fast and accurately. Figure 4.2 shows the *G*-ratio variation as a function of the grinding force due to the threshold of material removal and the breakdown of grinding wheel. This variation suggests that if the grinding force can

Fig. 5.18 Closed-loop step response under PI control

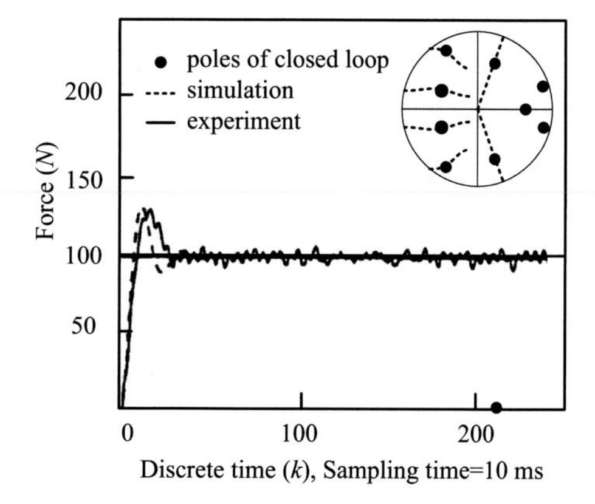

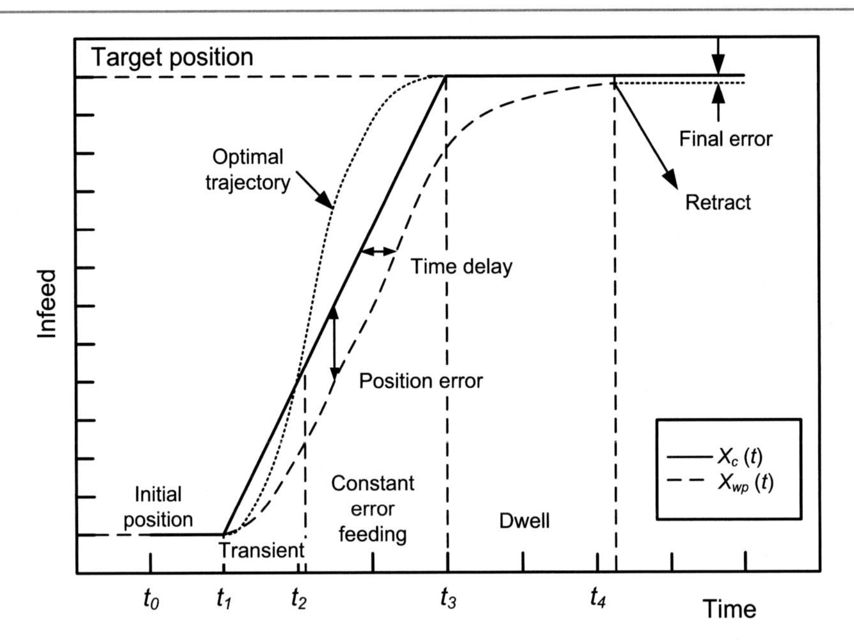

Fig. 5.19 The infeed motion of a cylindrical grinding wheel and its time delay and positioning error

be maintained at around the wheel breakdown point, with the implementation of feedback control, the grinding process can achieve optimal productivity.

The fundamental force equation for the grinding process was introduced by Hahn, who showed the relation between the thrust force, f_N, and the material removal rate, Z_w, through the parameter Λ_w, which depends on the material and process conditions.

$$Z_w = \Lambda_w (f_N - f_{Th}) \approx \Lambda_w f_N \tag{5.25}$$

The threshold force, f_{Th}, is the maximal force due to plowing and rubbing while no material removal takes place, and it can be negligible because it is much smaller than the normal force.

In traverse cylindrical grinding, Z_w can be calculated from the geometrical relationship of:

$$Z_w = \pi d_w a_p v'_{trav} \tag{5.26}$$

where d_w is the workpiece diameter, a_p is the radial depth of cut, and v'_{trav} is the traverse velocity.

Combining Eqs. (5.25) and (5.26) gives the relationship between the force and velocity:

$$F_N = \frac{Z_w}{\Lambda_w} = \frac{\pi d_w a_p}{\Lambda_w} v'_{trav} = K_{FV} v'_{trav} \tag{5.27}$$

The parameter K_{FV} can be considered constant if the cutting depth is the same for each pass and the piece diameter is almost constant (this can be assumed when the diameter is much greater than the cutting depth in each pass). The v'_{trav} represents the velocity at the contact point between the wheel and the piece. Since usually this velocity is not available from direct measurement, it is necessary to establish a model that relates the velocity of this point with the measured velocity by the encoder.

When grinding forces are present, the machine, tool, and workpiece react with a deflection δ, thereby introducing a difference between the real tool position (x_p) and the position measured by the encoder (x_e). This deflection can be described by a first order response between the applied force and the deflection as

$$F_N = K\delta + B\frac{\partial \delta}{\partial t} = K(x_e - x_p) + B\frac{\partial (x_e - x_p)}{\partial t} \quad (5.28)$$

where K and B represent the spring and the damping coefficients respectively.

The transfer function between the force and the measure velocity is found by taking the Laplace transform of Eqs. (5.27) and (5.28) and noting that $V'_{trav} = X_p s$ and $V_{trav} = X_e s$:

$$F_N(s) = \frac{K_{FV}(K + Bs)}{K + (K_{FV} + B)s} V_{trav}(s) \quad (5.29)$$

The thrust force (f_N) is related to the tangential force (f_T) by the friction coefficient of μ:

$$f_N = \frac{f_T}{\mu} \quad (5.30)$$

For a constant wheel velocity, the power consumed by the motor wheel (p_w) and the tangential force are related as

$$p_w = F_T v_{tw} \quad (5.31)$$

Taking the Laplace transform of Eqs. (5.30) and (5.31) and combining these with Eq. (5.29), the transfer function between the power and the traverse velocity is

$$P_w(s) = v_{tw}\mu \frac{K_{FV}(K + Bs)}{K + (K_{FV} + B)s} V_{trav}(s) = \frac{a_0 + a_1 s}{b_0 + b_1 s} V_{trav}(s) \quad (5.32)$$

This is a first-order transfer function (one pole and one zero) with coefficients (a_i and b_i), which represent the dynamics between the power consumed by the motor wheel and the traverse velocity for cylindrical traverse grinding. The importance of Eq. (5.32) is to define a structure of the dynamic process model, of which the parameters are apparently machine dependent and still have to be identified experimentally.

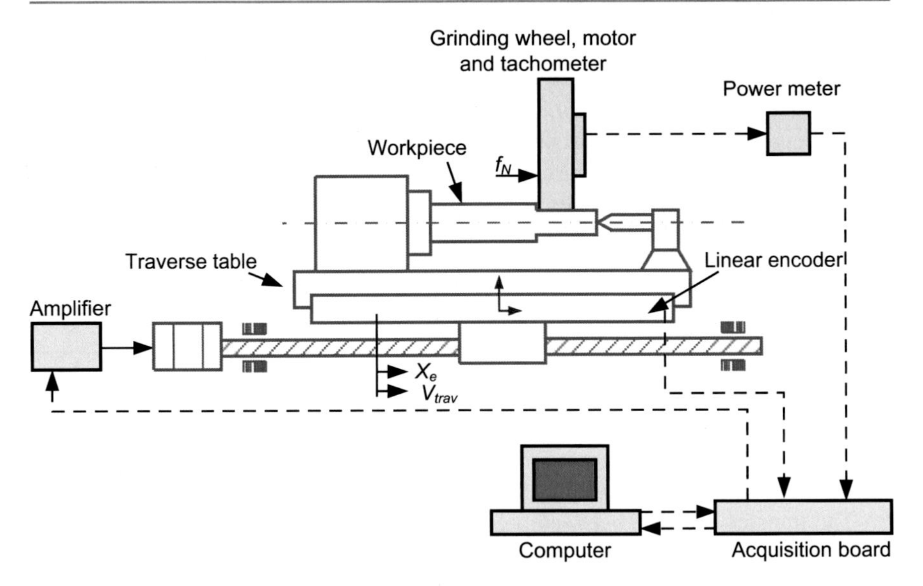

Fig. 5.20 Traverse grinding table and its control system components

A typical grinding machine with power feedback control is shown in Fig. 5.20. The control scheme is depicted in the block diagram in Fig. 5.21. Note that there are two loops of control: an outer loop for the power control and an inner loop for the velocity control. The inner loop is needed to overcome any table–motor dynamics so that the table closely follows the controlling commands from the power controller.

An example application is performed on a CNC cylindrical grinder with expanded sampling and control capability based on a microprocesser. A linear encoder, with a resolution of 0.5 μm was also attached to the table, to measure the position and calculate the table velocity. The power was measured by a Hall effect sensor with a full scale of 10 HP for a current network of 10 A. The controllers in the inner and outer loops are all of standard PI type, tuned to fulfill satisfactory dynamics requirements in terms of zero steady state error, settling time (<3 s), and overshoot (<15 %). The resulting response of power and velocity to a series of depth of cut variations is shown in Fig. 5.22. In this case the desirable power set point is 1400 W. Note that the table velocity automatically changes according to the depth of cut to maintain the constant level of power. If a traditional constant feed, instead of the power feedback, system were used, a conservative velocity of 2 mm/s would have been used throughout the pass and a 27 % loss of cycle time resulted.

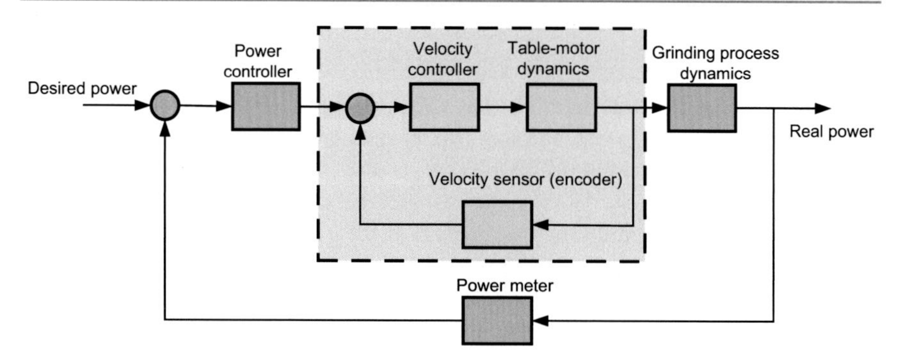

Fig. 5.21 Block diagram of the control system

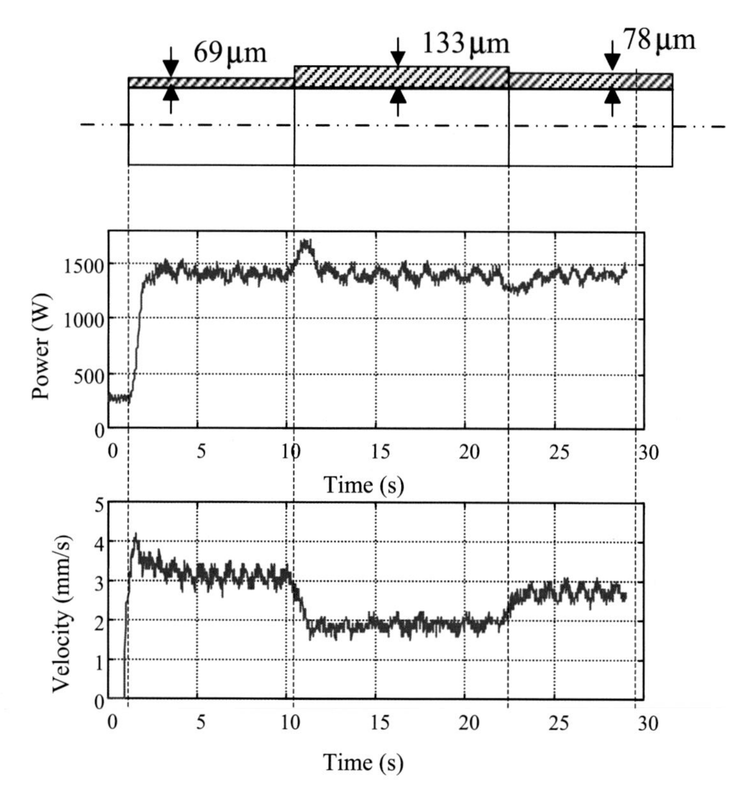

Fig. 5.22 Traverse velocity and power response for a reference power of 1400 W

Homework

1. Name two cutting tool materials that were not mentioned in the lectures notes. List their chemical compositions, critical temperatures, and industry applications. Also list your reference(s).
2. In a two-bearing main spindle, the distance between the bearings is a design parameter to give a minimum deflection at the cutting point. Identify such an optimal distance, l, for $a = 350$ mm, spindle diameter = 100 mm, $E = 21 \times 10^4$ N/mm^2, $I = 500$ cm^4, front (collar)-end bearing stiffness = 1000 N/μm, back (drive)-end-bearing stiffness = 600 N/μm. Restrict the distance l to between 50 mm and 500 mm.
3. Draw schematics of the square, T, and V types of box way configurations. Discuss the advantages and limitations of each configuration relative to one another.
4. In addition to resolvers, encoders, and scales, what are the sensors commonly used for the purpose of machine tool feedback control? Name three. Specify their measured quantity, physical principles, operation range, and accuracy. Also list your references(s).
5. In the classroom lecture we discussed an example of determining feed drive servomotor torque requirement. That example is based on the non-cutting case with table moving at rapid traverse velocity. Estimate the torque requirement for the cutting case of 1.5 mm/s feed rate and 8000 N feed force and 2000 N vertical force. The same servo rise time specification remains.
6. Simulate the force response of turning process under PI control with a reference force level of 175 N. Use the same process dynamics and the PI controller structure given in Eqs. (5.22) and (5.23) with $bp_s = 0.1$ and sampling time as 0.01 s. Present two time plots of force response, similar to Fig. 5.17 (without the root locus on the upper right), for two cases under different sets of PI controller constants: one stable force response case and the other unstable force response case. You choose the controller constants for these cases, and they do not have to be optimized in any way. You may perform this simulation by programming your own dynamic simulation calculation—one value at a time step, starting from zero force initial condition—using any language/spreadsheet. Alternatively, you may use any dynamic simulation package (for example, the Simulink in Matlab) of your choice.

6 Machine Tool Accuracy and Metrology

Accuracy is one of the most critical indicators of a machine tool's value. The error induced in the manufacturing of a machine tool can account for as much as 75 % of the total part of the error formed under the use of that machine tool. Most of the components rejected due to out of tolerance are not the results of operator's ignorance or material imperfection but because of the distortions, wear, or initial inaccuracies of the machines. The precise basing, corrosion resistance, thermal conductivity, antifrictional and frictional properties, strength and rigidity, and many others as discussed in Chap. 5 collectively determine the final accuracy of the machine tool.

On the other hand, increasing the accuracy of machines, as a general rule, not only improves their quality considerably but also increases the cost. Therefore, the establishment of the optimum accuracy level is a technical and economical optimization problem. In general, the requirements for the accuracy of machines are largely determined by their designation and the requirements of the principal indexes of their operation.

The objectives of this chapter are to introduce the basic indexes of machine tool accuracy, to discuss the common measurement method (both in terms of metrology technique and equipment) for the determination of accuracy, and to develop the ability to calculate machine accuracy in various machining configurations.

6.1 Acceptable Accuracy of Machine Tools

The accuracy of machine tool typically is referenced to certain inspection criteria and permissible deviations based upon international or national standards or acceptance recommendations from professional associations. At the international level, the International Organization for Standardization (ISO) defines the standardization for its member nations. At the national level in Germany, there are standards suggested by Machine Tool Standards Committee (NWM), Verein Deutscher Ingenieure (VDI), etc.

S. Y. Liang, A. J. Shih, *Analysis of Machining and Machine Tools*,
DOI 10.1007/978-1-4899-7645-1_6

Table 6.1 Allowed deviation (μm) of N class machines ($L=1000$ mm or $D=32$ mm)

Category	Objective	Key parameter	Max δ
Surface positioning of tool and work	Planeness	Largest length, L	36
	Linearity	L	22
	Roundness	Largest diameter, D	20
Motion of working element	Linearity	L	22
	Radial play of rotation	D	16
	Axial play of rotation	D	16
Tool or work relative to guides	Parallelism	L	28
	Eccentricity	D	40

According to a standard published by *Standartizatsiya* in Russia, for example, machine tools with respect to accuracy are divided into five classes: N (normal accuracy), E (elevated accuracy), H (high accuracy), A (especially high accuracy), and C (ultimately high accuracy). Some examples of required accuracy of N class machines are given in Table 6.1. The allowed deviations for the principal indexes of accuracy when mapping from one class to the next higher class are arranged according to a geometrical series with a base of 1.6. In other words, the allowed deviations for the E class machines are 1.6 times less than for the N class accuracy. The deviations are 1.6^2 times less for H class machines. The key parameter also plays a role as the magnification factor in the square root sense. That is, the allowable deviation of machines of class N of other sizes (with other values of L's and D's) can be estimated from $\delta'=\delta\sqrt{L'/1000}$ or $\delta'=\delta\sqrt{D'/32}$.

Example: The tool parallelism for an H class machine, 850 mm long and 45 mm in diameter, is $\delta' = (28)\left(\sqrt{\dfrac{850}{1000}}\right)/1.6^2$ μm.

6.2 Machine Tool Accuracy

The accuracy of a machine tool is generally defined with respect to the errors involved in the movement of positioning axes. The errors are defined as the geometric deviations of actual positions from their ideal ones of a machine component along its axis of motion. These deviations can be caused by static or dynamic sources. Possible static sources include component weight and guide imperfection. Dynamic sources can be attributed to cutting forces, motion, acceleration, etc. The general definition of machine tool accuracy does not attempt to treat these sources differently, while the final positioning errors that determine the part performance are of ultimate interest.

Discussion in this chapter will use the milling machine as an example since it represents a generic configuration. The knowledge derived from the milling machine can be readily extended to other cutting configuration by simplification. Our

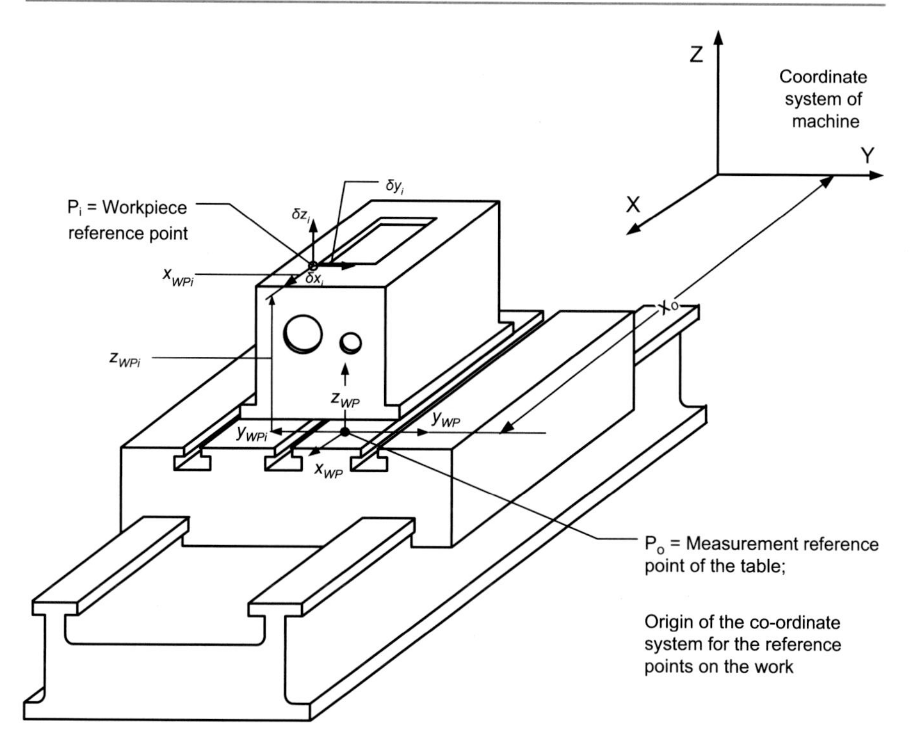

Fig. 6.1 Machine coordinate and component coordinate

interest in the error analysis is focused on the exact location of a certain point, which is often the cutting point, on the workpiece attached to the machining table. The exact location of such a point, referred to as the reference point hereafter, can be defined based on two independent coordinate systems. One coordinate system is the so-called component coordinate, which gives the position of the reference point with respect to the machining table. The other coordinate system is the machine coordinate that gives the position of machining table with respect to the earth. If only 1-D motion in the feed (X) direction is considered, these two coordinate systems can be shown in Fig. 6.1. The $(x_{WPi}, y_{WPi}, z_{WPi})$ are the component coordinates of the reference point P_i and (x, y, z) are the machine coordinates. In this case, $y=z=0$ and $x=x_0$, where x_0 is the nominal feed.

Let us consider the error associated with the machining table first. Due to the possible distortion of the machining table, for any point P_o in the plane of the machining table (fed along the X axis only), three translatory errors are identified in the orthogonal axis directions, as shown in Fig. 6.2. These are $\Delta x_T(x_0) = x - x_0 =$ positional error in the feed direction, $\Delta y_T(x_0) = y(x_0) =$ straightness error normal to the feed direction, and $\Delta z_T(x_0) = z(x_0) =$straightness error normal to the feed direction as well. Note that these errors are defined with respect to three axes of feed, crossfeed, and height that are nominally perpendicular to one another, that is, the three axes are perfectly square.

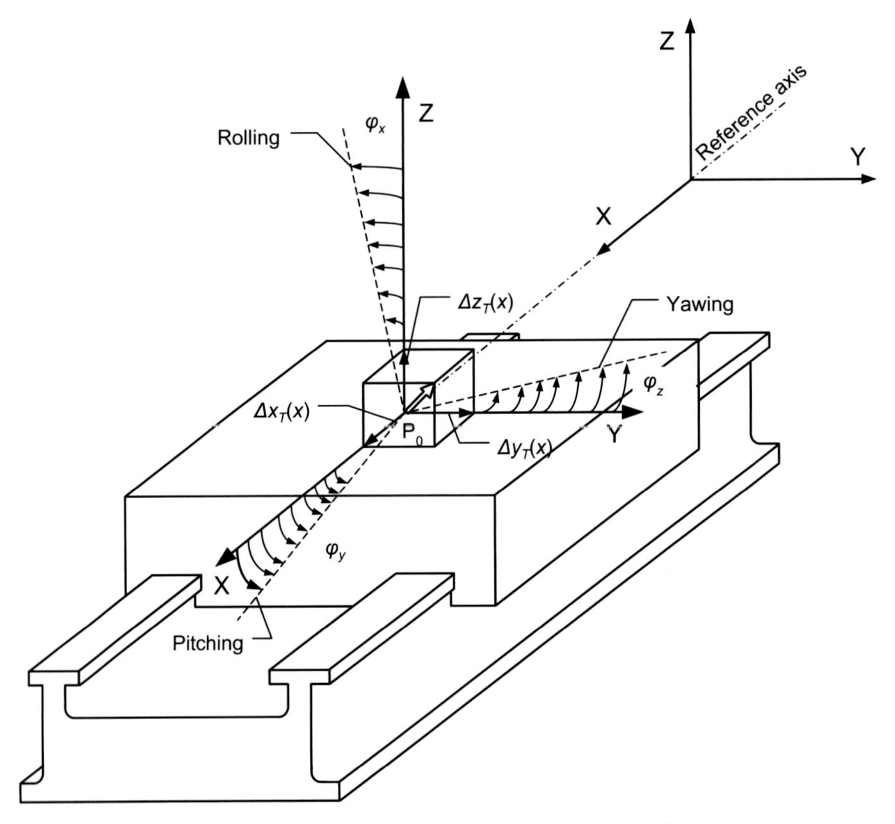

Fig. 6.2 Roll, pitch, and yaw associated with feed motion in the X direction

To describe the position of the reference point P_i with respect to the point P_o, the rotational errors φ_x, φ_y, and φ_z as shown in Fig. 6.2 must be considered in addition to the translatory errors. These are φ_x = rotational error about the X axis, also known as rolling, φ_y = rotational error about the Y axis, pitching, and φ_z = rotational error about the Z axis, yawing. Note that these rotational errors are associated with the rotation of axes, which are mutually perpendicular to one another.

The above concept is sufficient to define the errors of every point on the workpiece associated with the 1-D feed motion as:

$$\begin{Bmatrix} \Delta x_i\left(x_o, X_{WPi}\right) \\ \Delta y_i\left(x_o, X_{WPi}\right) \\ \Delta z_i\left(x_o, X_{WPi}\right) \end{Bmatrix} = \begin{Bmatrix} \Delta x_T\left(x_o\right) \\ \Delta y_T\left(x_o\right) \\ \Delta z_T\left(x_o\right) \end{Bmatrix} + \begin{bmatrix} 0 & -\varphi_z\left(x_o\right) & \varphi_y\left(x_o\right) \\ \varphi_z\left(x_o\right) & 0 & -\varphi_x\left(x_o\right) \\ -\varphi_y\left(x_o\right) & \varphi_x\left(x_o\right) & 0 \end{bmatrix} \begin{Bmatrix} x_{WPi} \\ y_{WPi} \\ z_{WPi} \end{Bmatrix}, \quad (6.1)$$

where Δx_i, Δy_i, Δz_i = the error of point P_i relative to the earth, as functions of both machine coordinate and component coordinate; $X_{WPi} = \left\{x_{WPi}, y_{WPi}, z_{WPi}\right\}$, the component coordinate; $\left\{\Delta x_T, \Delta y_T, \Delta z_T\right\}$ = translatory errors; and $\left\{\varphi_x, \varphi_y, \varphi_z\right\}$ = rotational errors in terms of rolling, pitching, and yawing.

Example

The translatory and rotational (rolling, pitching, and yawing) errors associated with a 1-D feed motion x_o are given in the following figures. Estimate the greatest error associated with a point with component coordinate of (50 mm, 0 mm, 10 mm) in terms of either straightness or positional accuracy.

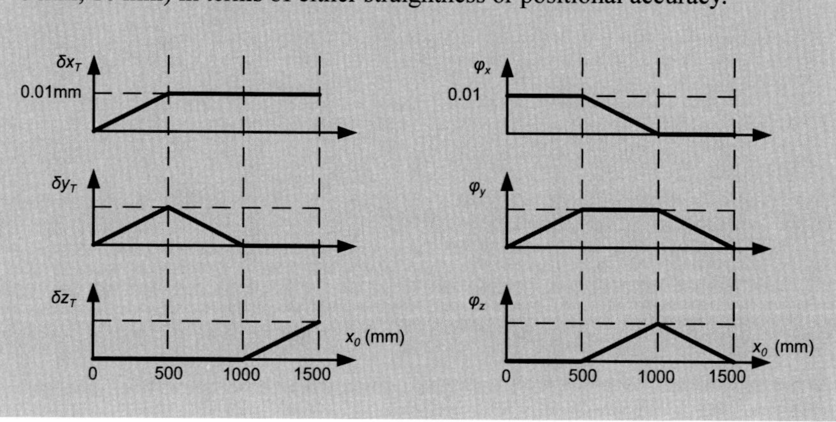

Solution:

$$\begin{Bmatrix} \Delta x_i \\ \Delta y_i \\ \Delta z_i \end{Bmatrix} = \begin{Bmatrix} \Delta x_T(x_o) \\ \Delta y_T(x_o) \\ \Delta z_T(x_o) \end{Bmatrix} + \begin{bmatrix} 0 & -\varphi_z(x_o) & \varphi_y(x_o) \\ \varphi_z(x_o) & 0 & -\varphi_x(x_o) \\ -\varphi_y(x_o) & \varphi_x(x_o) & 0 \end{bmatrix} \begin{Bmatrix} 50 \\ 0 \\ 10 \end{Bmatrix}$$

$$= \begin{Bmatrix} \Delta x_T \\ \Delta y_T \\ \Delta z_T \end{Bmatrix} + \begin{Bmatrix} 10\varphi_y \\ 50\varphi_z - 10\varphi_x \\ -50\varphi_y \end{Bmatrix}$$

With the total error:

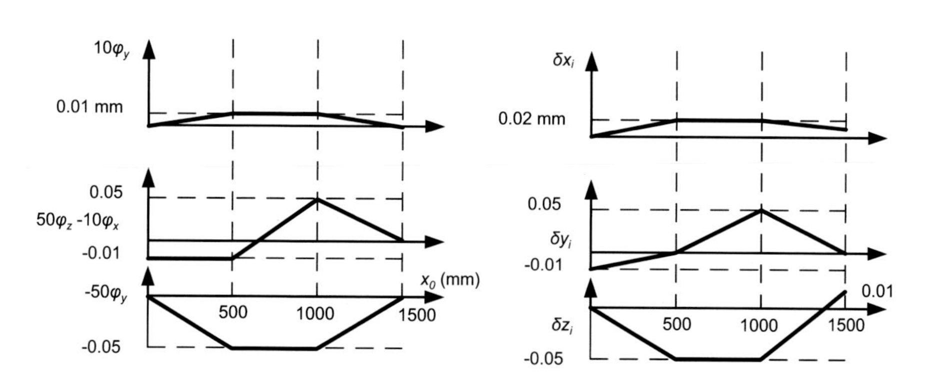

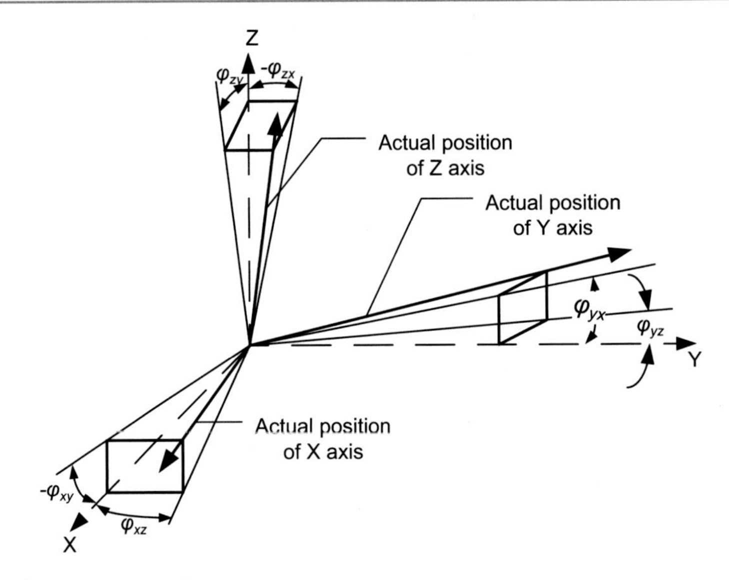

Fig. 6.3 Squareness errors between axes

Equation 6.1 involves a basic assumption—the three machining table motion axes, namely the X, Y, and Z axes, are perfectly square. In reality, when multi-axis motion occurs, the three axes can have independent angular deflections leading to additional position errors. When such independent angular deflections take place, the three axes are not necessarily perpendicular to one another; and in a general error analysis, the angular deflections of the axes with respect to one another must also be taken into consideration.

In Fig. 6.3, ϕ_{xy} and ϕ_{xz} represent the angular errors of the X axis relative to the Y and Z axes, respectively. Other errors are defined in a similar way. In such a representation, the axes are assumed to exhibit an angular deflection only to the first order, that is, they are always straight, although angular rotation between them is possible. If table distortion and the rolling, pitching, and yawing in the component coordinate system are not considered, the position errors of the reference point P_o with respect to the earth due to angular deflections are then directly proportional to its traveling distance along the machine coordinate axes.

$$\begin{Bmatrix} \Delta x_{\measuredangle}(\mathrm{X}) \\ \Delta y_{\measuredangle}(\mathrm{X}) \\ \Delta z_{\measuredangle}(\mathrm{X}) \end{Bmatrix} = \begin{bmatrix} 0 & -\varphi_{yz} & \varphi_{zy} \\ \varphi_{xz} & 0 & -\varphi_{zx} \\ -\varphi_{xy} & \varphi_{yx} & 0 \end{bmatrix} \begin{Bmatrix} x_o \\ y_o \\ z_o \end{Bmatrix}, \tag{6.2}$$

where $\Delta x_{\measuredangle}$, $\Delta y_{\measuredangle}$, $\Delta z_{\measuredangle}$ are errors due to angular deflections and $\mathrm{X} = \{x_o, y_o, z_o\}$ in making a multi-axis motion. These angular deflections have nothing to do with the component coordinate, as they merely reflect the positioning error of point P_o (in Fig. 6.1) as caused by the multi-axis feed.

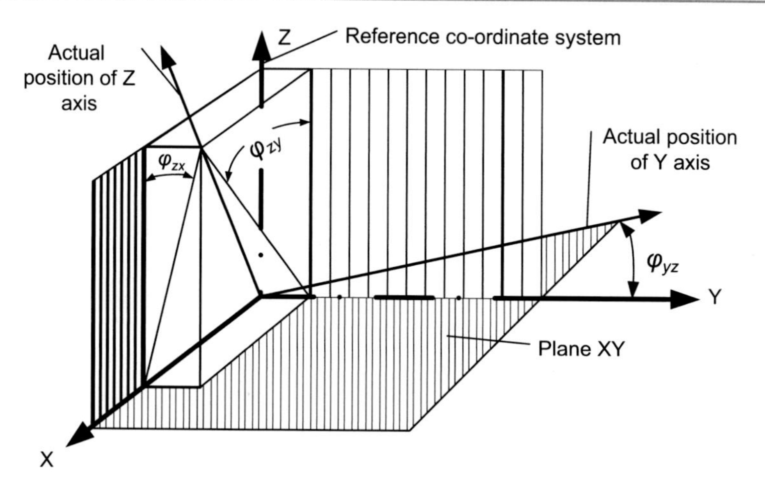

Fig. 6.4 A degenerated case of squareness error representation

Equation (6.2) can be simplified if one of the axes, say the feed (X axis), is chosen as the datum axis and the machining table (XY plane) the datum plane, as shown in Fig. 6.4. Then φ_{xz}, φ_{xy}, φ_{yx} and therefore $\Delta z_{\measuredangle}$ are zero and the other errors are:

$$\begin{Bmatrix} \Delta x_{\measuredangle}(\mathrm{X}) \\ \Delta y_{\measuredangle}(\mathrm{X}) \end{Bmatrix} = \begin{bmatrix} -\varphi_{yz} & \varphi_{zy} \\ 0 & -\varphi_{zx} \end{bmatrix} \begin{Bmatrix} y_o \\ z_o \end{Bmatrix} \tag{6.3}$$

Generally speaking, in the case of multi-axis motion, the total position error of any point on the workpiece is the linear superposition of translatory errors (due to machining table distortion), rotational errors (due to rolling, pitching, and yawing), and angular errors (due to table axis nonsquareness). Therefore,

$$\begin{Bmatrix} \Delta x_i(\mathrm{X}, \mathrm{X}_{WPi}) \\ \Delta y_i(\mathrm{X}, \mathrm{X}_{WPi}) \\ \Delta z_i(\mathrm{X}, \mathrm{X}_{WPi}) \end{Bmatrix} = \begin{Bmatrix} \Delta x_T(\mathrm{X}) \\ \Delta y_T(\mathrm{X}) \\ \Delta z_T(\mathrm{X}) \end{Bmatrix} + \begin{bmatrix} 0 & -\varphi_z(\mathrm{X}) & \varphi_y(\mathrm{X}) \\ \varphi_z(\mathrm{X}) & 0 & -\varphi_x(\mathrm{X}) \\ -\varphi_y(\mathrm{X}) & \varphi_x(\mathrm{X}) & 0 \end{bmatrix} \begin{Bmatrix} x_{WPi} \\ y_{WPi} \\ z_{WPi} \end{Bmatrix} + \begin{bmatrix} 0 & -\varphi_{yz} & \varphi_{zy} \\ \varphi_{xz} & 0 & -\varphi_{zx} \\ -\varphi_{xy} & \varphi_{yx} & 0 \end{bmatrix} \begin{Bmatrix} x_o \\ y_o \\ z_o \end{Bmatrix}, \tag{6.4}$$

where, again, $\mathrm{X}_{WPi} = \{x_{WPi}, y_{WPi}, z_{WPi}\}$ is the component coordinate and $\mathrm{X} = \{x_o, y_o, z_o\}$ the multi-axis feed motion.

The errors of the machine components are responsible for workpiece dimensional error, which develops during the cutting process on the machine. The dimension of a workpiece can refer to its height, width, length, taper, corner radius, etc. Suppose a dimension is defined as the linear distance between two points A and B, the dimensional errors are:

$$\begin{Bmatrix} \Delta x_{\overline{AB}} \\ \Delta y_{\overline{AB}} \\ \Delta z_{\overline{AB}} \end{Bmatrix} = \begin{Bmatrix} \Delta x_A\left(\mathrm{X}_{oA}, \mathrm{X}_{WPA}\right) \\ \Delta y_A\left(\mathrm{X}_{oA}, \mathrm{X}_{WPA}\right) \\ \Delta z_A\left(\mathrm{X}_{oA}, \mathrm{X}_{WPA}\right) \end{Bmatrix} - \begin{Bmatrix} \Delta x_B\left(\mathrm{X}_{oB}, \mathrm{X}_{WPB}\right) \\ \Delta y_B\left(\mathrm{X}_{oB}, \mathrm{X}_{WPB}\right) \\ \Delta z_B\left(\mathrm{X}_{oB}, \mathrm{X}_{WPB}\right) \end{Bmatrix}, \tag{6.5}$$

where X_{oA} is the position of reference point P_o at the time when point A is cut by a tool (or measured by a probe), and X_{oB} is the position of reference point P_o at the time when point B is cut or measured. Note that if all items in Eq. (6.4) are known then the position errors at point A $(\Delta x_A, \Delta y_A, \Delta z_A)$ and those at point B can be calculated; thus, the dimensional error can be estimated for given tool path. This is to assume that the cutting force and machining table dynamics do not play a role in the generation of part errors. Also note that the dimensional error given by Eq. (6.5) is a 3-D vector, sufficient to provide information needed for tolerance stacking calculations.

6.3 Random Error and Tolerance Template

In addition to the deterministic errors described above, random errors may also be present. Such random or nonsystematic errors may vary from test to test and hence statistical treatment is usually applied to quantify their effects.

Figure 6.5 illustrates an example of measurement results over the length of travel along one axis in both forward and backward directions. Next to the mean-value curves, several readings (at least 5) under the same conditions are shown in terms of $3s$, where s is the standard deviation. The difference between the forward and the backward mean values is referred to as the hysteresis, backlash, or dead zone, which is introduced by the play in the transmission components and the friction forces.

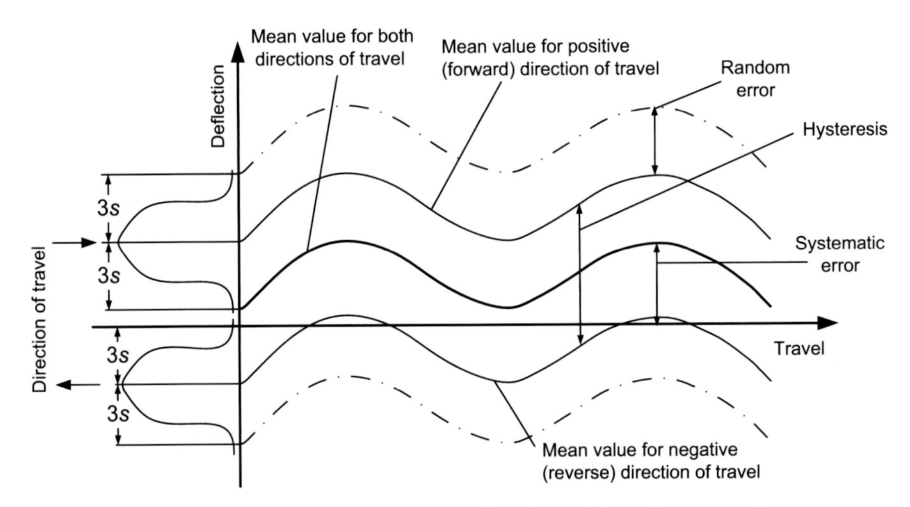

Fig. 6.5 Single-axis travel positional error in both directions with random uncertainty

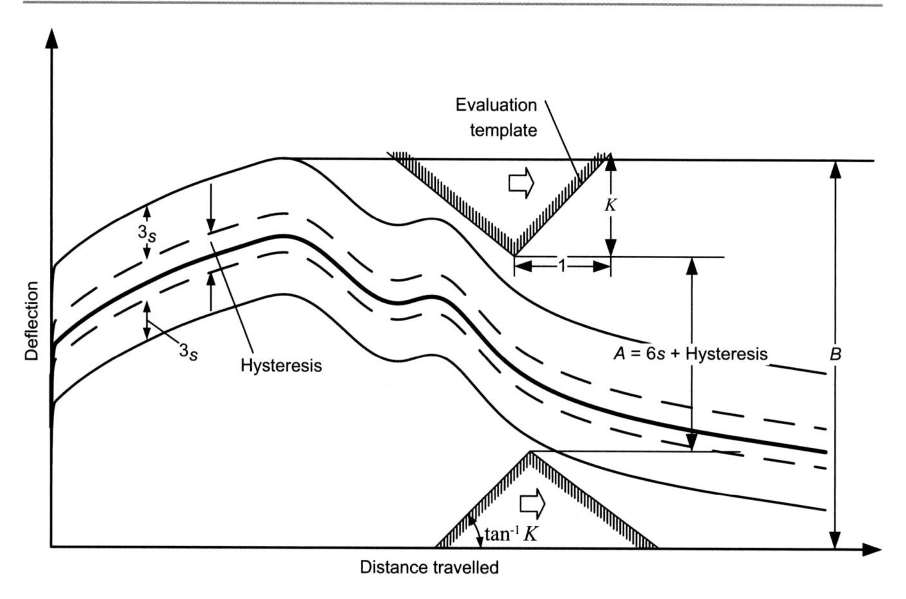

Fig. 6.6 Positional error evaluated against a tolerance template

The errors measured are often evaluated against specified tolerances, as suggested by the National Machine Tool Builders Association with the use of tolerance templates. Figure 6.6 shows such a double-sided template, if shifted along the error field, that should always be able to include the entire error. The two characteristic parameters of the template are the narrowest gap A and the slope K. Typically, these tolerance attributes may be $A = 10$ μm and $K = 10$ μm/m for high-quality machine tools. For the machine accuracy to be acceptable, any two points, x_1 and x_2, (both are nominal feed positions), along a positioning path should satisfy:

$$|\Delta x(x_1) - \Delta x(x_2)| < A + K|x_1 - x_2| \tag{6.6}$$

This standard can be extended to feed motion in multi-axis situations.

6.4 Machine Tool Metrology

The metrology of machine tools deals with the measurement of the translatory and rotational terms that are responsible for position errors for any point on the workpiece. These terms are represented quantitatively in Eq. (6.4) as the 3-D translatory errors δX_T, the rotational error (roll, pitch, and yaw) of the component coordinate system $\{\varphi_x, \varphi_y, \varphi_z\}$, and angular (squareness) errors between axes $\{\varphi_{xz}, \varphi_{xy}, \varphi_{yz}, \varphi_{yx}, \varphi_{zx}, \varphi_{zy}\}$. Once these characteristics are measured, the position errors can be represented as functions of component coordinates; thus the entire machine tool error distribution can be mapped out.

When the measurement of these geometric characteristics is carried out, the machine is not normally subject to cutting forces and workpiece weight. However, these factors can affect the static, dynamic, and thermal machine characteristics as well as its geometric behaviors during actual machining operation. Therefore, it is desirable to conduct machine tool metrology under defined and representative loading conditions.

6.4.1 Measurement of Straightness Errors

Straightness errors are the terms of $\Delta y_T(x_o)$ and $\Delta z_T(x_o)$ in the case of a single-axis x_o feed motion for example. The use of a focused light beam and a four-quadrant photodiode is common for straightness measurement as shown in Fig. 6.7. The light beam (e.g., laser beam) is arranged parallel to the feed path being examined. The beam shines on a photodiode, which registers the intensity of the incident light on each of its four quadrants separately. The beam source is fixed, while the photodiode is moved with the table. Movement in the plane perpendicular to the direction of travel produces variations of light intensity in the four quadrants of the photodiode. By careful evaluation of the signal intensity obtained from the quadrants, the deviations from a straight-line motion can be determined in both vertical and horizontal directions.

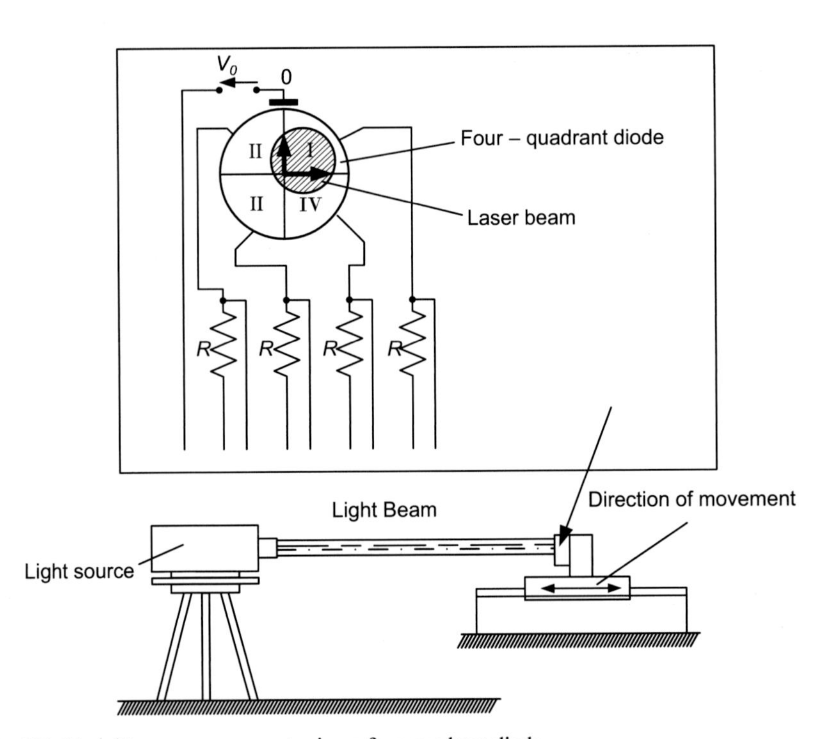

Fig. 6.7 Straightness measurement using a four-quadrant diode

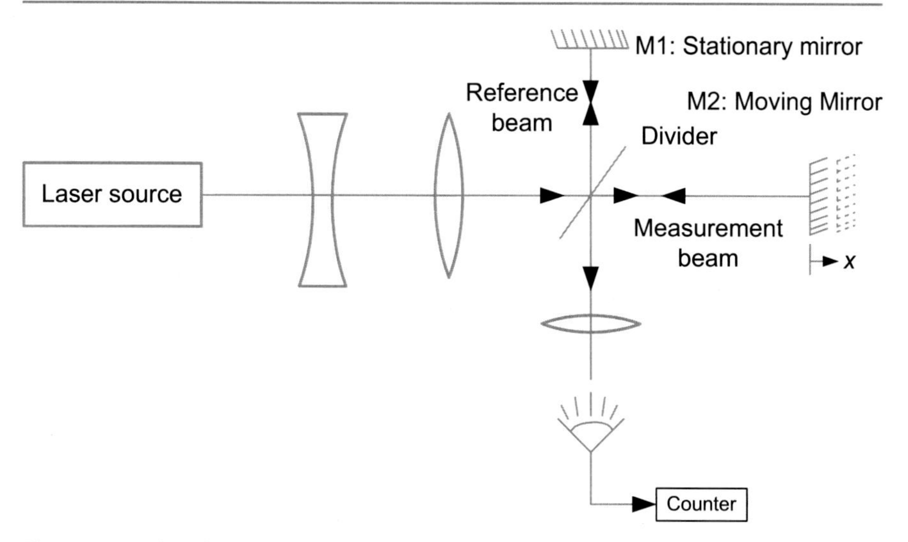

Fig. 6.8 Laser interferometry

6.4.2 Measurement of Feed Positional Error

The feed positional error is defined as the magnitude of the difference between the theoretical position and the actual position. This would be the term of $\Delta x_T(x_o)$ in the case of a single-axis x_o feed motion for example. This is affected by a number of factors such as the resolution and accuracy of the linear measuring system, elasticity of the drive components, the inertia forces when speeding up and braking, friction and stick-slip in the guideways, table or slide movement resulting from clamping after position, etc. The performance of the CNC control and the competence of machine operation can also affect the accuracy.

Laser interferometry is often used to measure feed positional error. Figure 6.8 shows the basic principle of interferometry with a light beam of a single wavelength and phase impinges upon a beam divider, where it is split into two beams: reference and measurement. The former is reflected off a stationary mirror M1 and the latter off a mirror M2 attached to the moving machine feed axis along the same line of beam traveling. Both the reference and the measurement beams reach the photodetector eventually but at different phases due to the difference in the travel distance. If the travel distance difference is an integer multiple of the beam wavelength, the resulting light intensity at the photodetector shows a maximum. If the distance difference is an integer and a half of the wavelength, the intensity is at minimum as shown in Fig. 6.9. Therefore, the theoretical resolution of the interferometer measurement of displacement is half of the beam wavelength (e.g., 0.316 μm for helium-neon laser)—although the environmental conditions such as air pressure, temperature, humidity, and carbon dioxide content can also influence the accuracy. As the M2 location changes, the light intensity at the photodetector goes through a number of minimum and maximum cycles. Once the number of cycles is counted,

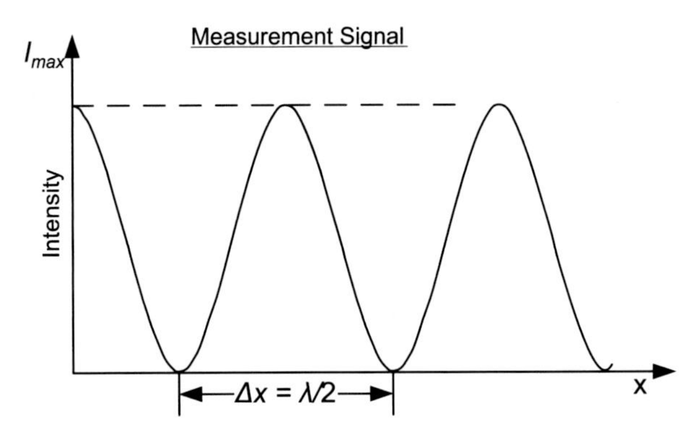

Fig. 6.9 Intensity versus displacement curve of interferometer measurement

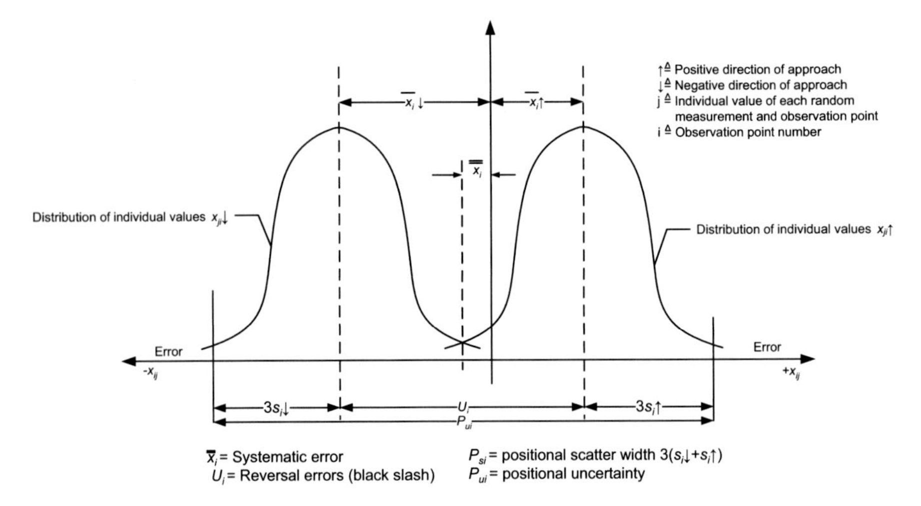

Fig. 6.10 Statistical distribution of error measurement data

the total distance of M2 travel can be determined. Note that this technology is good for the measurement of motion and displacement; it is useless for the static measurement of distances.

The positioning accuracy is a very important index of the machine tool quality and therefore its value. The reporting of such index has to be a standardized acceptance test procedure without ambiguity. For example, VDI (Verein Deutscher Ingenieure—Association of German Engineers)/DGQ (Deutsche Gesellschaft Fur Qualitat—German Association for Quality) sections 3441–3445 mandate that in an acceptance test, the slides have to be moved repeatedly to a number of reference positions. This is done for each axis from both directions of travel, where their positional errors are determined with the aid of laser interferometers. Then the following data, as illustrated in Fig. 6.10, are calculated:

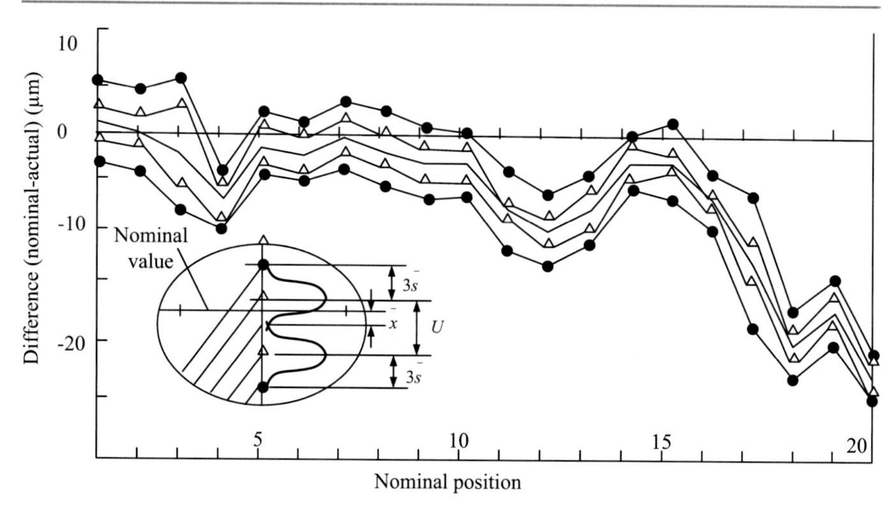

Fig. 6.11 Example evaluation of statistical data in forward and backward directions

- Reversal errors (backlash): $U_i = \left|\bar{x}_{i\uparrow} - \bar{x}_{i\downarrow}\right|$, where $\bar{x}_{i\uparrow}$ is the mean value (accounting for all repeated tests) for positive direction of approach at location i and $\bar{x}_{i\downarrow}$ is for negative direction.
- Scatter width: $P_{Si} = 6\bar{s}_i = 6\left(\frac{s_{i\uparrow} + s_{i\downarrow}}{2}\right)$, where $s_{i\uparrow}$(standard deviation) $= \sqrt{\frac{1}{n-1}\sum_{j=1}^{n}\left(x_{ji\uparrow} - \bar{x}_{i\uparrow}\right)}$ and n is the total number of test repetitions.
- Positioning uncertainty $P_{Ui} = U_i + P_{Si}$.
- Mean positional error$=\bar{\bar{x}}_i = \frac{\bar{x}_{i\uparrow} + \bar{x}_{i\downarrow}}{2}$.

Figure 6.11 shows an example of a characteristic chart for the positioning accuracy of one axis of a machining center, which was obtained as a result of a standard acceptance test. Faults in the feed mechanism can be diagnosed as: (a) large amount of backlash in position 3 possibly due to a fault in the feed shaft and (b) steadily increasing error between the positions 15 and 20 possibly due to a lead-screw pitch error.

Example

Use the first five data points in Fig. 6.11 to estimate the positioning uncertainty P_{ui} and mean positional error $\bar{\bar{x}}_i$.

Solution:

The first five points in Fig. 6.11 are ($i=1, 2, 3, 4, 5$):

$$3s_{i\uparrow} = [2.17,\ 2.25,\ 2.41,\ 1.04,\ 1.12].$$

$$3s_{i\downarrow} = [2.31,\ 2.46,\ 2.53,\ 1.13,\ 1.18].$$

$$\bar{x}_{i\uparrow} = [-0.64,\ -1.21,\ -4.51,\ -7.76,\ -2.66].$$

$$\bar{x}_{i\downarrow} = [2.41,\ 1.69,\ 2.50,\ -4.43,\ 0.72].$$

$$U_i = \left|\bar{x}_{i\uparrow} - \bar{x}_{i\downarrow}\right| = [3.05,\ 2.90,\ 7.01,\ 3.33,\ 3.38].$$

$$U_i = \left|\bar{x}_{i\uparrow} - \bar{x}_{i\downarrow}\right| = [3.05,\ 2.90,\ 7.01,\ 3.33,\ 3.38].$$

so, $P_{Ui} = U_i + P_{Si} = U_i + 3s_{i\uparrow} + 3s_{i\downarrow} = [7.53, 7.61,\ 11.95,\ 5.50,\ 5.68].$

$$\bar{\bar{x}}_i = \frac{\bar{x}_{i\uparrow} + \bar{x}_{i\downarrow}}{2} = \left[0.885,\ 0.24,\ -1.005,\ -6.095,\ -0.97\right].$$

6.4.3 Measurement of Roll, Pitch, and Yaw

To measure φ_x, φ_y, and φ_z in Eq. (6.4), suitable optical devices can be used with the combination of two linear measurements. Figure 6.12 shows the basic configuration based on two laser interferometers. Figure 6.13 gives the set up for measuring the pitch motion, while the roll and the yaw can be acquired in a similar way. Note that the two moving mirror setup provides the magnitude of angular motion, but it does not provide the sign of it. To know the sign, two sets of system identical to Fig. 6.8, which involves stationary mirrors, have to be used.

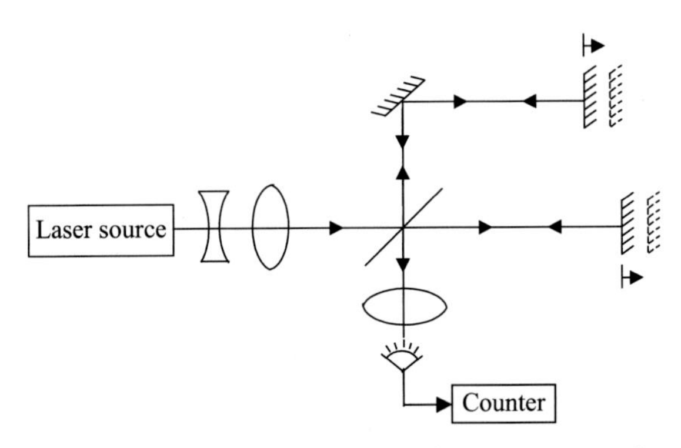

Fig. 6.12 Measure of angular position based on twin laser interferometer reflectors

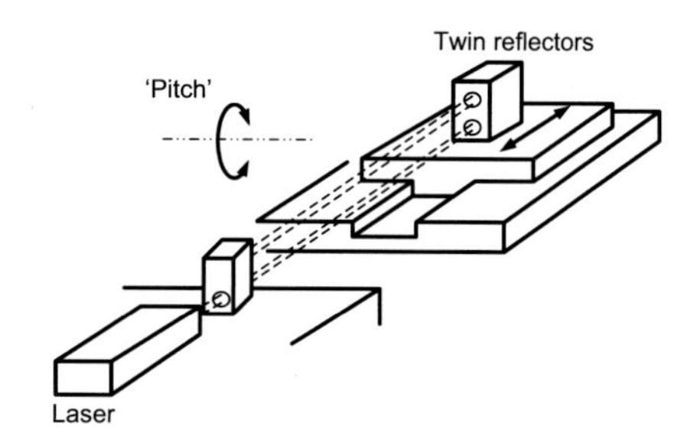

Fig. 6.13 Measure of pitch motion using twin reflectors

6.4.4 Measurement of Squareness Errors

The squareness errors between any two axes are the terms of φ_{yz}, φ_{zy}, φ_{xz}, φ_{zx}, φ_{xy}, and φ_{yx} in Eq. (6.4). They can be measured with a four-quadrant photodiode in a two-stage procedure. Figure 6.14 shows the stages involved using Z and Y axes as an example.

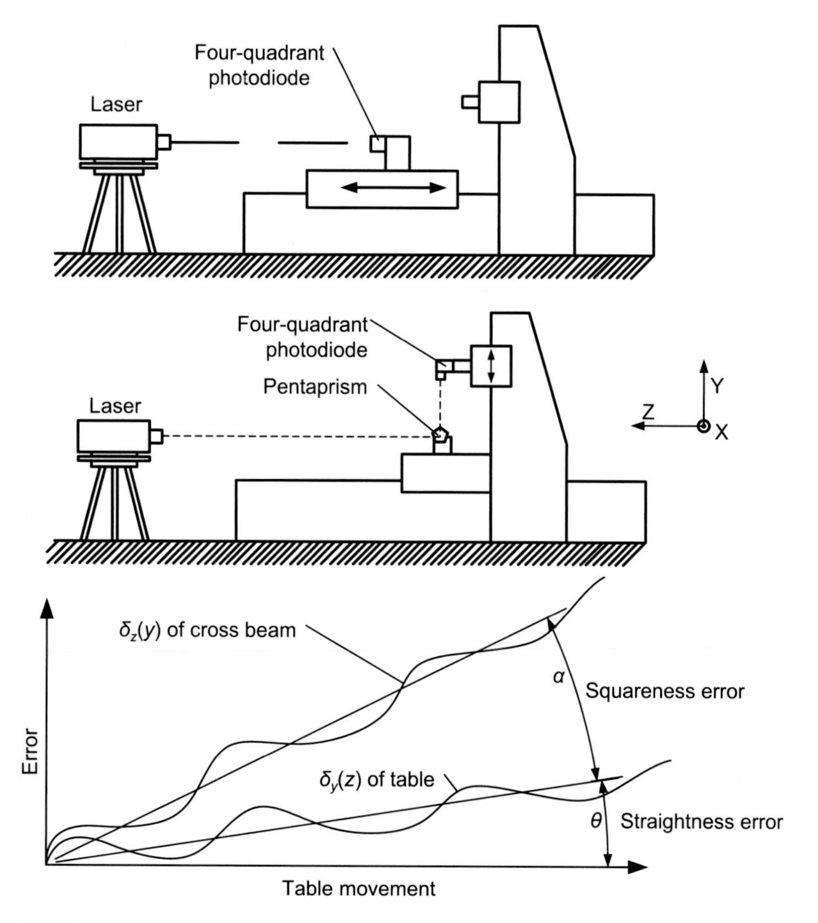

Fig. 6.14 Squareness measurement between two axes of movement

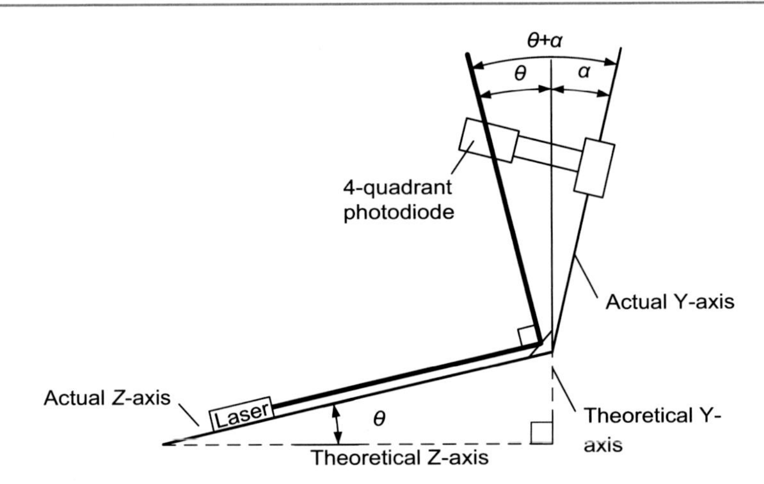

Fig. 6.15 Geometric relationship between θ and α in the squareness measurement

At first, the straightness of the Z direction table movement is measured. To do this, a four-quadrant photodiode is fixed to the table. The readings taken are recorded on the error versus movement plot as the lower curve $\Delta y(z)$. Note that the mean straight line produces an angle θ with the abscissa.

At a second stage, a mirror is positioned on the Z table, which is now kept stationary. A beam diverted at an angle of 90° by the mirror is in the movement direction of Y. A four-quadrant photodiode is fitted on to the tool holder. The straightness of travel $\Delta z(y)$ of the Y movement is now measured and plotted in Fig. 6.14 as the upper curve. The mean straight line of the two curves produces an angle α between each other, which represents the error in squareness between the two machine axes. This angle α is thus φ_{yx}. Figure 6.15 depicts the relationship between θ and α. Note that the squareness error is defined with respect to the theoretical Z axis, but not the actual Z axis.

Homework

1. For a point "i," given $\begin{Bmatrix} \Delta x_T \\ \Delta y_T \\ \Delta z_T \end{Bmatrix}(X_o) = \begin{Bmatrix} 0 \\ 0.01 \\ 0.01 \end{Bmatrix}$ mm, $\begin{Bmatrix} x_{WPi} \\ y_{WPi} \\ z_{WPi} \end{Bmatrix} = \begin{Bmatrix} 0 \\ 10 \\ 50 \end{Bmatrix}$ mm,

$$\varphi_x(X_o) = \varphi_y(X_o) = 0.001 \text{ rad}, \ \varphi_z(X_o) = 0, \ \begin{Bmatrix} x_o \\ y_o \\ z_o \end{Bmatrix} = \begin{Bmatrix} 100 \\ 50 \\ 50 \end{Bmatrix} \text{mm},$$

$$\varphi_{xy} = \varphi_{yx} = \varphi_{yz} = \varphi_{zy} = \varphi_{zx} = \varphi_{xz} = 0.001 \text{ rad, calculate } \begin{Bmatrix} \Delta x_i \\ \Delta y_i \\ \Delta z_i \end{Bmatrix}(X_o, X_{WPi}).$$

2. For another point "j" in addition to the point "i" in Problem 1, given

$$\begin{Bmatrix} \Delta x_T \\ \Delta y_T \\ \Delta z_T \end{Bmatrix}(X_o) = \begin{Bmatrix} 0.01 \\ 0 \\ 0.01 \end{Bmatrix} \text{mm} \begin{Bmatrix} x_{WPi} \\ y_{WPi} \\ z_{WPi} \end{Bmatrix} = \begin{Bmatrix} 50 \\ 10 \\ 50 \end{Bmatrix} \text{mm}, \quad \varphi_x(X_o) = \varphi_z(X_o) = 0.001 \text{ rad},$$

$$\varphi_y(X_o) = 0, \begin{Bmatrix} x_o \\ y_o \\ z_o \end{Bmatrix} = \begin{Bmatrix} 150 \\ 100 \\ 100 \end{Bmatrix} \text{mm}, \; \varphi_{xy} = \varphi_{yx} = \varphi_{yz} = \varphi_{zy} = \varphi_{zx} = \varphi_{xz} = 0.001 \text{ rad},$$

calculate the part dimensional error of segment ij.

3. Given $A = 0.05$ mm, $b = 20$ mm, and T (or K) $= 0.0001$ for the tolerance template as shown below, estimate the allowable standard deviation s for acceptance of the machine used in Problems 1 and 2 considering the positioning errors between the two specified points. Note that all three orthogonal feed directions (X, Y, and Z) need to meet the template standard. Shown in the figure below is the tolerance examination in one (X) of the feed directions. Assume zero backlash in all directions.

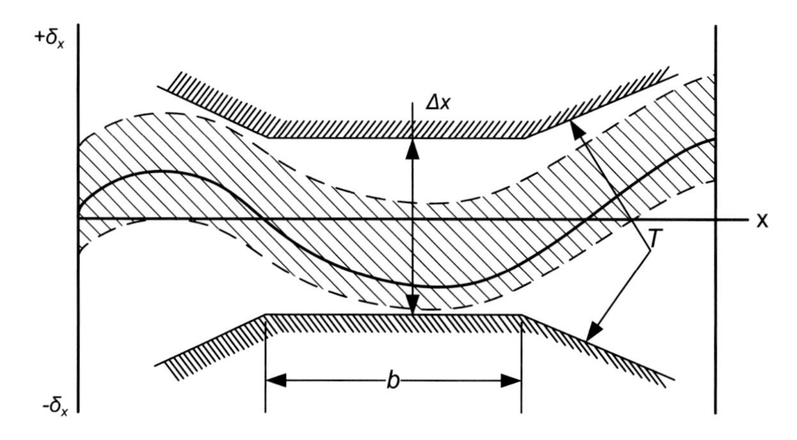

4. Design a system to measure the parallelism between two axes of motion in the above figure. Draw a schematic on the figure and explain the operation principles.

Mechanics of Machining

7

7.1 Orthogonal Cutting Model

In the study of machining mechanics, the basic chip formation process by the removal of material from a workpiece by the action of a wedge-shaped cutting edge of the tool is considered. The objective of this study is to provide an analytical basis which relates cutting forces, tool stresses, and temperature, etc., to the cutting conditions such as the cutting speed, size of cut, cutting tool geometry, and the material properties of the workpiece and the tool. It would be then possible to determine such practically important factors as the cutting power from the forces, dynamic stability from the tool–work interface condition, and the effective life of the tool from the tool temperatures and stresses. There has been a lot of work devoted into the use of empirical relations derived based on experimental measurements for this purpose. Unfortunately, experimental measurements are quite time consuming and costly, especially in view of the constant introduction of new workpiece and tool materials. In addition, a small deviation from the nominal composition of a workpiece material can cause large changes in its machining characteristics. To reduce the experimental work, it is clear that more fundamental relationships than purely empirical ones are needed. The goal of this chapter on the mechanics of machining is to provide these.

In both experimental and analytical investigations of chip formation, it has been common to consider the relatively simple case of orthogonal machining. Shown in Fig. 7.1 is a tool with a plane cutting face and a single, straight cutting edge, which is normal to the cutting velocity v (velocity of work relative to tool), removing a layer of work material of uniform depth t (commonly termed as the undeformed chip thickness) and width b. These dimensions are related to a single-point turning process as shown in Fig. 2.1 of Chap. 2 by $t = a_c = f \sin j_r$ and $b = a_p / \sin j_r$ in which j_r is the major cutting-edge angle of the tool. The geometry of the cutting edge can in this case be defined by its width, which in general is greater than the width of cut, and by the two angles α and θ. The angle α between the tool cutting

S. Y. Liang, A. J. Shih, *Analysis of Machining and Machine Tools*,
DOI 10.1007/978-1-4899-7645-1_7

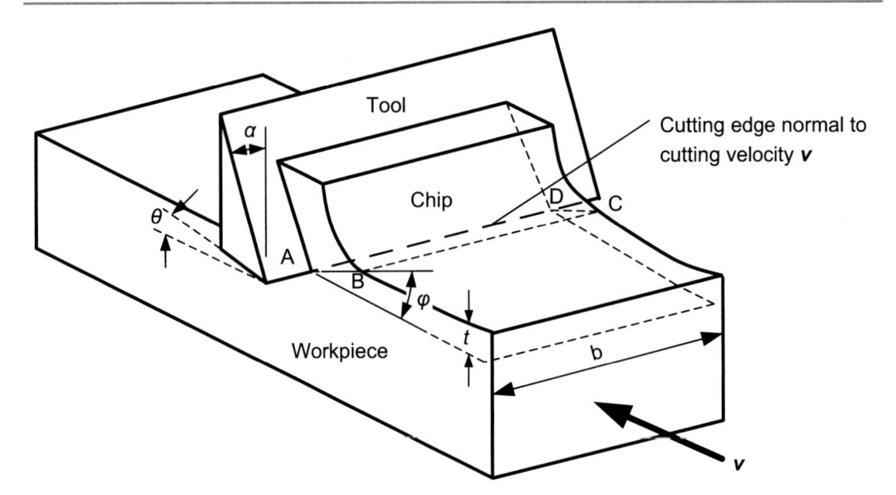

Fig. 7.1 An orthogonal machining model

face and the normal to the cutting velocity $\boldsymbol{v}$ is termed the rake angle; it is measured positive as shown in the figure and negative if the tool rake leans toward the chip. Experiments show that the rake angle has a profound effect on the chip formation process and hence on cutting forces, power, etc. The angle θ between the clearance face of the tool and the machined work surface is termed the clearance angle. It has been shown that the clearance angle can influence the rate of clearance face wear; therefore the value of the clearance angle is determined by cutting-edge strength considerations and the need for the tool to clear the machined surface.

A model of this sort is two-dimensional, and to be closely approximated by this ideal model, a machining process should satisfy the following assumptions: perfectly sharp tool, plane strain, constant depth of cut, constant and uniform cutting velocity, continuous chip formation, no built-up edge on tool, uniform shear and normal stress along shear plane and tool. Two of the earliest researchers to employ this model were Ernest and Merchant, who suggested that there is a border separating the deformed and undeformed regions in the workpiece. This border can be reasonably represented by a flat plane, called the shear plane, or primary deformation plane. As seen in Fig. 7.1, the shear plane, ABCD, making a shear angle φ with the cutting velocity direction divides the workpiece from the chip. The material below this plane is undeformed, while the chip above it has been deformed by a concentrated shearing action.

7.2 Forces in Orthogonal Cutting

Understanding the forces and power involved in cutting is important for the following reasons: Power and torque requirements have to be determined before a drive motor of suitable capacity can be designed; appropriate design of the machine tool structure that avoids excessive distortion of the machine tool elements and

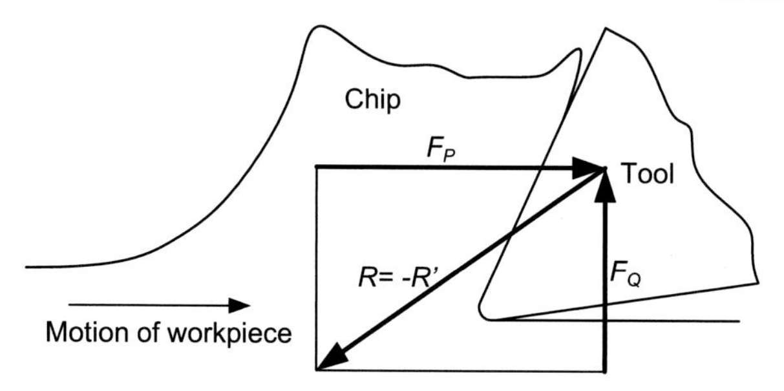

Fig. 7.2 Cutting and thrust components of resultant tool force

maintains desired tolerances for the machined part is possible only if cutting force data are available; whether the workpiece can withstand the cutting forces without excessive distortion has to be determined in advance; and the cutting forces are usually the main sources for forced vibration between the work and the tool, and therefore their magnitudes are important for the analysis of machine tool dynamics.

In orthogonal cutting, the resultant force R applied to the chip by the tool lies in a plane normal to the tool cutting edge as shown in Fig. 7.2. This force is usually determined, in experimental work, from the measurement of two orthogonal components: one in the direction of cutting velocity (referred to as the cutting force F_P), the other normal to the direction of cutting velocity (referred to as the thrust force F_Q). The accurate measurement of these two components of the resultant tool force has been commonly available with the use of dynamometers that measure the piezoelectric charges of quartz or the deflections (or strain) in elements supporting the cutting tools. The two components F_P and F_Q may be used to calculate many important variables in the process of chip formation.

Note that the resultant force can also be resolved into two components on the tool face: a friction force, F_C, along the tool–chip interface, and a normal force, N_C, perpendicular to the interface. These two force components are related by the friction coefficient at the tool–chip interface. Designating the friction coefficient as μ,

$$\mu = \tan\beta = \frac{F_C}{N_C}, \tag{7.1}$$

where β is referred to as the friction angle.

The resultant force is balanced by an equal and opposite force along the shear plane and is resolved into a shear force, F_S, and a normal force N_S. Figure 7.3 shows a free body diagram of the chip being produced. Using the resultant force R as the diameter, a force circle can be constructed such that all the cutting forces begin and terminate on the circle as shown in Fig. 7.4. The construction of this circle facilitates the derivation of the following relationships:

$$F_S = F_P\cos\varphi - F_Q\sin\varphi \tag{7.2}$$

$$N_S = F_P\sin\varphi + F_Q\cos\varphi = F_S\tan(\varphi + \beta - \alpha) \tag{7.3}$$

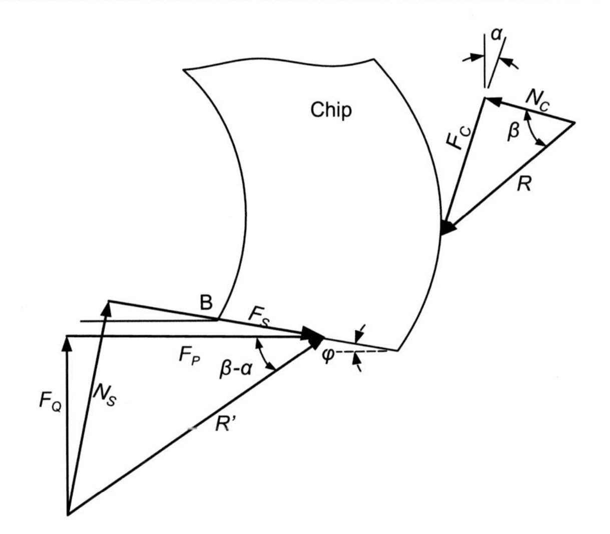

Fig. 7.3 Free body diagram of a chip

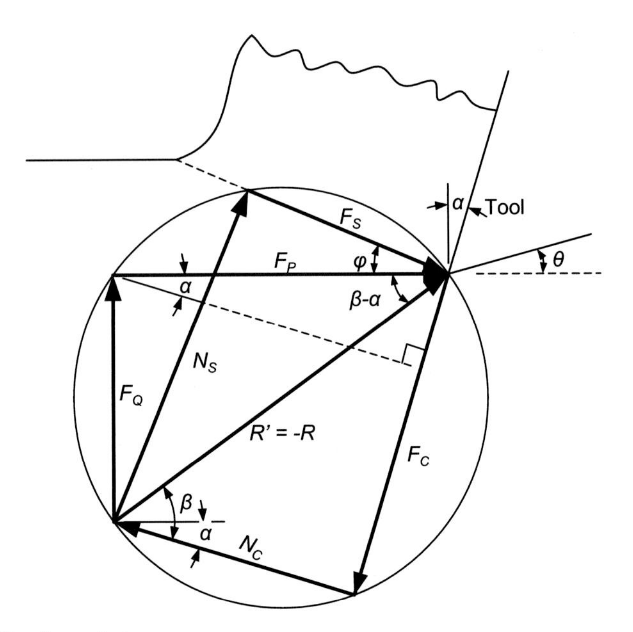

Fig. 7.4 Cutting force circle

$$F_C = -F_P \sin \alpha - F_Q \cos \alpha \tag{7.4}$$

$$N_C = -F_P \cos \alpha + F_Q \sin \alpha = F_C \cot \beta. \tag{7.5}$$

The friction coefficient is thus:

$$\mu = \frac{F_C}{N_C} = \frac{F_P \sin\alpha + F_Q \cos\alpha}{F_P \cos\alpha - F_Q \sin\alpha} = \frac{F_Q + F_P \tan\alpha}{F_P - F_Q \tan\alpha}. \tag{7.6}$$

7.3 Stresses

From the shear plane components of cutting force the mean shear and normal stresses on the shear plane can be determined:

$$\tau = \frac{F_S}{A_S}, \tag{7.7}$$

$$\sigma = \frac{N_S}{A_S},$$

where A_S is the area of the shear plane. For a width of cut b and a depth of cut t,

$$A_S = \frac{bt}{\sin\varphi}. \tag{7.8}$$

Therefore,

$$\tau = \frac{(F_P \cos\varphi - F_Q \sin\varphi)\sin\varphi}{bt} \tag{7.9}$$

and

$$\sigma = \frac{N_S}{A_S} = \frac{(F_P \sin\varphi + F_Q \cos\varphi)\sin\varphi}{bt}. \tag{7.10}$$

The shear stress τ in Eq. (7.9) is the apparent shear strength of the material that has to be overcome to shear the workpiece and form the chip along the shear plane. Experimental work has shown that the apparent shear strength calculated in this way remains constant for a given work material over a wide variety of cutting conditions. It has been observed, however, that at small undeformed chip thickness the shear stress τ decreases with increasing undeformed chip thickness. This exception to the constancy of τ can be explained by the existence of a constant plowing force at the tool cutting edge, details of which will be discussed in Chap. 8.

7.4 Shear Angle

The relationships discussed in the previous section cannot be effectively utilized without the knowledge of the shear angle φ. Basically, there are three methods with which the shear angle can be estimated: direct measurement off a photomicrograph,

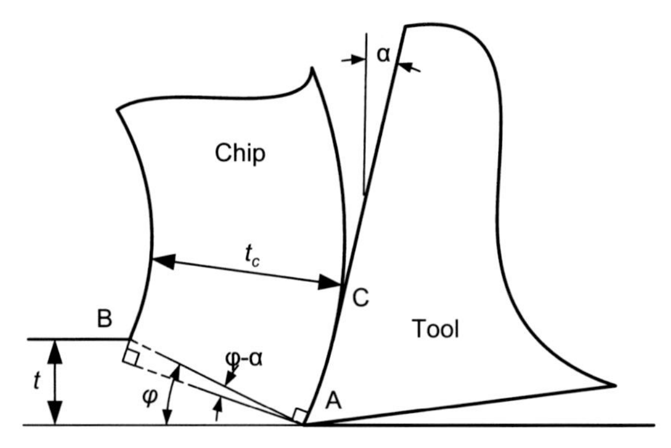

Fig. 7.5 Shear angle and cutting ratio

calculation from the cutting ratio, and the prediction based on analytical shear angle solutions.

A photomicrograph can be acquired by quick stop and removal of the tool during the course of a cutting operation followed by the polishing and etching of the cross section of the chip and work, then the microscope photographing of the etched section. This procedure is usually difficult and requires specialty equipment.

The cutting ratio is defined as the ratio of the axial depth of cut (undeformed chip thickness) to the chip thickness, that is,

$$r = \frac{t}{t_C}. \tag{7.11}$$

From Fig. 7.5, it can be seen that

$$AB = \frac{t_C}{\cos(\varphi - \alpha)} = \frac{t}{\sin\varphi}. \tag{7.12}$$

Therefore,

$$\varphi = \tan^{-1}\frac{r\cos\alpha}{1 - r\sin\alpha}. \tag{7.13}$$

Experimentally, it has been found that in cutting there is no change in the work material density. In addition, width of cut remains constant according to the plane strain assumption in the orthogonal machining model, which is valid for the case of large width of cut in comparison to the depth of cut. Considering the volume conservation in plasticity in conjunction with the above gives

$$r = \frac{t}{t_c} = \frac{l_c}{l} = \frac{v_c}{v}. \tag{7.14}$$

Equations (7.13) and (7.14) suggest a rather straightforward methods of estimating the shear angle using the measurement of the chip length l_c and the corresponding work length l, as well as the chip velocity v_c.

The first complete theoretical analysis resulting in a so-called shear angle solution was presented by Ernst and Merchant. In their analysis, the chip is assumed to behave as a rigid body held in equilibrium by the action of the forces transmitted across the chip–tool interface and across the shear plane. They argued that the shear angle φ would take up such a value as to reduce the work done in cutting to a minimum. Since, for a given cutting velocity, the work done in cutting was proportional to F_P, it was necessary for F_P to be a minimum. From the force circle, it can be shown that

$$F_P = \tau \frac{bt}{\sin\varphi} \frac{\cos(\beta - \alpha)}{\cos(\varphi + \beta - \alpha)}. \tag{7.15}$$

Differentiating the above expression with respect to φ and equating it to zero gives

$$\varphi = \frac{\pi}{4} - \frac{\beta}{2} + \frac{\alpha}{2}. \tag{7.16}$$

Example
A new material was turned on a lathe using a cutter of 90 ° major cutting-edge angle and 10 ° rake angle. It was measured that the cutting force is 400 lb, and the thrust force is 150 lb. If the same material is to be turned with a 15 ° cutter while all other cutting conditions remain unchanged, could you estimate the forces involved?

Solution:

Based on Eq. (7.6), $\mu = \dfrac{F_Q + F_P \tan\alpha}{F_P - F_Q \tan\alpha} = \dfrac{150 + 400(\tan 10°)}{400 - 150(\tan 10°)} = 0.59.$

Based on Eq. (7.16), $\varphi = \dfrac{\pi}{4} - \dfrac{\beta}{2} + \dfrac{\alpha}{2} = \dfrac{\pi}{4} - \dfrac{\tan^{-1} 0.59}{2} + \dfrac{10°}{2} = 34.7°.$

Equation (7.9) states that

$$\tau = \frac{(F_P \cos\varphi - F_Q \sin\varphi)\sin\varphi}{bt}$$
$$= \frac{(400\cos 34.7° - 150\sin 34.7°)\sin 34.7°}{bt} = \frac{138.6}{bt}.$$

Assuming the friction angle does not change with the rake angle,

$$\varphi' = \frac{\pi}{4} - \frac{\beta}{2} + \frac{\alpha'}{2} = \frac{\pi}{4} - \frac{\tan^{-1} 0.59}{2} + \frac{15}{2} = 37.2°,$$

and assuming the shear stress does not change with the rake angle either,

$$\tau' bt = \left(F'_P \cos 37.2^\circ - F'_Q \sin 37.2^\circ\right)\sin 37.2^\circ = \tau bt = 138.6 \text{ lb},$$

and

$$\mu' = \frac{F'_Q + F'_P \tan 15^\circ}{F'_P + F'_Q \tan 15^\circ} = \mu = 0.59.$$

From the above two equations, it can be found that $F'_P = 364$ lb and $F'_Q = 101$ lb. Note that the assumptions made, especially regarding the constant τ, may not be true. Nevertheless, it is generally observed in actual machining that a larger rake angle often indeed reduces the amount of cutting force in view of the smaller amount of plastic deformation required to achieve the same material removal rate.

Example

An engine lathe has a 5 HP motor at a spindle speed of 2400 rpm. A cylindrical workpiece of 5 in. diameter is being turned on this lathe by a carbide tool with 90° major cutting-edge angle and 10° rake angle. The workpiece material is medium carbon steel of 15 kpsi shear strength. No lubrication was used in this process, and the friction coefficient at the interface of chip and cutter rake is 0.4. While the feed and the radial depth of cut are freely adjustable, what is the maximum achievable material removal rate at the spindle speed of 2400 rpm? Assume that the power required to feed is rather negligible compared to that required to spin, therefore HP = (tangential cutting force, in lb) × π × (workpiece diameter, in ft) × (spindle rpm) / 33,000. Also note that the removal rate in turning is MRR = (feed) × (radial depth of cut) × (spindle rpm) × π × (workpiece diameter).

Solution:

The maximum available cutting force F_P is limited by the maximum HP:

$$\text{HP} = \frac{F_P \pi d(\text{rpm})}{33{,}000} \Rightarrow 5 = \frac{F_P \pi \left(\frac{5}{12}\right)(2400)}{33{,}000} \Rightarrow F_P = 52.5 \text{ lb}$$

$$\beta = \tan^{-1} 0.4 = 21.8^\circ \text{ and } \varphi = (90^\circ - \beta + \alpha)/2 = 39.1^\circ$$

Since $F_Q = F_P \tan(\beta - \alpha)$,

$$bt = \frac{\sin\varphi(F_P\cos\varphi - F_Q\sin\varphi)}{\tau} = \frac{\sin\varphi F_P(\cos\varphi - \tan(\beta-\alpha)\sin\varphi)}{\tau}$$

$$= \frac{\sin 39.1°(52.5)\left[\cos 39.1° - \tan(21.8° - 10°)\sin 39.1°\right]}{15\times10^3}$$

$$= 1.4\times10^{-3}\ \text{in}^2,$$

and the maximum achievable $\text{MRR} = bt\,(\text{rpm})\,\pi d = 1.4\times10^{-3}(2{,}400)\,\pi\,(5) = 52.8\ \text{in}^3/\text{min}$.

Merchant found that the theory as given by Eq. (7.16) agreed well with experimental results obtained when cutting synthetic plastics but collapsed when compared to data for steel machined with a sintered carbide tool. It should be noted that in differentiating Eq. (7.15), it was assumed that τ would be independent of φ. On reconsidering these assumptions, Merchant decided to include in a new theory the relationship

$$\tau = \tau_o + K\sigma, \tag{7.17}$$

which states that the shear strength of the material τ increases linearly with the increase in normal compressive stress, σ, on the shear plane as shown in Fig. 7.6. This relationship has been supported in experimental observations, especially with polycrystalline metals. From the force circle it is derived that

$$\frac{\tau}{\sigma} = \frac{F_S}{N_S} = \cot(\varphi + \beta - \alpha). \tag{7.18}$$

Combining Eqs. (7.17) and (7.18) leads to:

$$\tau = \frac{\tau_o}{1 - K\tan(\varphi + \beta - \alpha)}. \tag{7.19}$$

Substituting τ into Eq. (7.15) gives

$$F_P = \frac{\tau_o bt\cos(\beta - \alpha)}{\sin\varphi\cos(\varphi + \beta - \alpha)[1 - K\tan(\varphi + \beta - \alpha)]}. \tag{7.20}$$

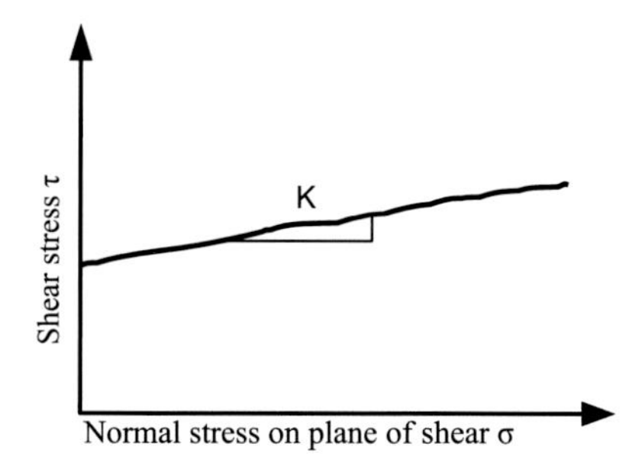

Fig. 7.6 Shear stress versus normal stress assumed in Merchant's second theory

It is now assumed that K and τ_o are constants for the particular work material and that b and t are constants for the cutting operation. Thus, Eq. (7.20) may be differentiated to give the value of φ,

$$2\varphi + \beta - \alpha = \cot^{-1} K. \tag{7.21}$$

Lee and Shaffer approached the solution of shear angle in orthogonal cutting using the slip line method in plasticity theory. In dealing with slip line problems, some basic assumptions are involved: The material is rigid plastic, the behavior of the material is strain-rate insensitive, cutting temperature effect is negligible, and the inertia effect resulting from chip acceleration is negligible.

In the solution of a problem, a slip line field can be constructed to consist of two orthogonal families of lines called the slip lines, indicating, at each point in the plastic zone, the two orthogonal directions of maximum shear stress. The slip line proposed by Lee and Shaffer for the orthogonal cutting of a continuous chip is shown in Fig. 7.7. This field assumes that all the deformation takes place on the shear plane. Consideration is given, however, to the manner in which the cutting forces applied by the tool are transmitted to the shear plane through a triangular plastic zone ABC on the chip.

If the boundaries of this triangular zone are considered, it is clear that the shear plane A must give the direction of one set of slip lines since the maximum shear stress must occur along the shear plane. Also, since no force acts on the chip after it has passed through the boundary B, no stresses can be transmitted across this boundary. Thus, B can be regarded as a free surface, and since the directions of maximum shear stress always meet a free surface at $\pi/4$, the angle between A and B must be equal to $\pi/4$. Finally, assuming that the stresses acting at the chip–tool interface are uniform (in a lot of cases this assumption may not be true), the principal stresses at the boundary D will meet this boundary at the friction angle β due to the fact that the resultant R has no shear component. Since directions of maximum shear stress lie at $\pi/4$ to the directions of principal stress, the angle η is given by $(\pi/4) - \beta$.

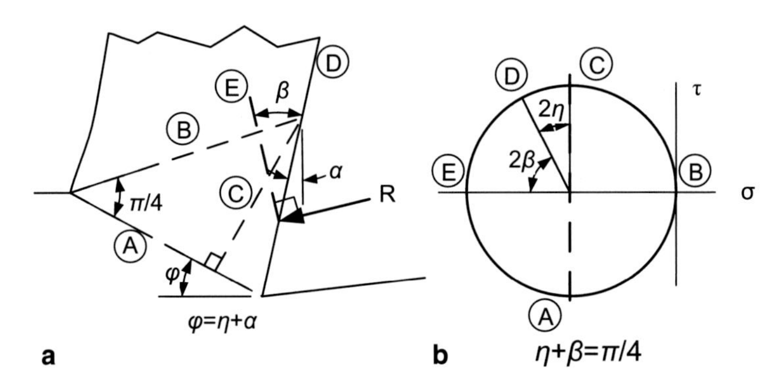

Fig. 7.7 Cutting force circle

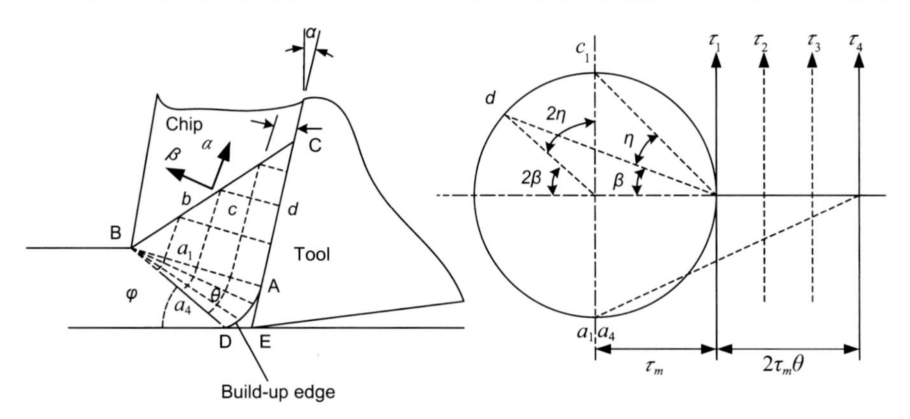

Fig. 7.8 Shear angle and cutting ratio

It now follows from Fig. 7.7 that

$$\varphi = \eta + \alpha = \frac{\pi}{4} - \beta + \alpha. \tag{7.22}$$

Lee and Shaffer realized that Eq. (7.22) could not apply where β is equal to $\pi/4$ and α is zero. They considered, however, that such conditions of high friction and low rake angle were just those conditions that lead to the formation of a built-up edge in practice. To support this point, a second solution was presented for the new geometry where a built-up edge is present on the tool face, in which

$$\varphi = \frac{\pi}{4} + \theta - \beta + \alpha. \tag{7.23}$$

Note that this solution is different from Eq. (7.22) by the size of built-up edge, as given by θ as seen in Fig. 7.8.

Example

A built-up edge is a major source of machined surface roughness, however, to identify the size of the built-up edge without interrupting the cutting process can be quite challenging since the built-up edge is often physically buried under the chip and therefore not observable from outside. A system as shown in the figure is now proposed to provide a nonintrusive estimate of built-up edge size during single-point turning operation. In this system, a vision camera, stationary with respect to the tool holder, measures the chip velocity while the dynamometer records the cutting forces. For a cylindrical workpiece, with a diameter of 2 in., being turned by a 90° major cutting-edge angle and 5° rake angle cutter at a feed of 0.002 ipr while the spindle speed is 1200 rpm, the chip velocity is measured to be 390 fpm, and the tangential cutting force and the thrust cutting force are found to be 160 lb and 50 lb, respectively. Estimate the built-up edge size (arc length AD in Fig. 7.8) as shown in the figure.

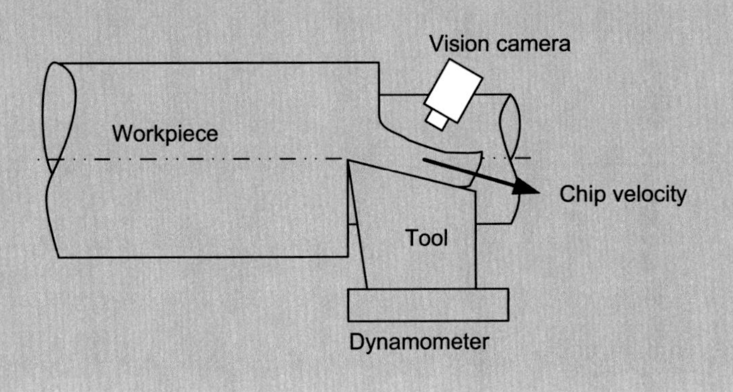

Solution:

Cutting velocity: $v = \dfrac{1200\pi(2)}{12} = 628$ fpm

Cutting ratio: $r = \dfrac{v_c}{v} = \dfrac{390}{628} = 0.62$

Shear angle: $\varphi = \tan^{-1}\dfrac{(0.62)\cos 5^{\circ}}{1-(0.62)\sin 5^{\circ}} = 33.2^{\circ}$

Friction angle: $\tan\beta = \dfrac{F_Q + F_P\tan\alpha}{F_P - F_Q\tan\alpha} = \dfrac{50+160\tan 5^{\circ}}{160-50\tan 5^{\circ}} \Rightarrow \beta = 22.4^{\circ}$

Modified Lee and Shaffer's shear angle solution:

$$\theta = \varphi - 45^{\circ} + \beta - \alpha = 33.2^{\circ} - 45^{\circ} + 22.4^{\circ} - 5^{\circ} = 5.6^{\circ}.$$

Built-up edge size: $b = \theta$ (in radian) $t = \dfrac{5.6\pi}{180^{\circ}}(0.02) = 1.93\times10^{-4}$ in.

The theories outlined above were compared with experimental results as seen in Fig. 7.9. Qualitatively speaking, experimental results show that a linear relationship exists between φ and $\beta-\alpha$ and that a decrease in $\beta-\alpha$ always results in an increase in φ. Thus, for a given rake angle, a decrease in the mean friction angle on the tool face results in an increase in the shear angle with a corresponding decrease in the area of shear. Since the mean shear strength of the work material in the shear zone remains constant, the force required to form the chip will be reduced. An increase in rake angle always results in an increase in the shear angle and hence a reduction in the cutting forces.

On the other hand, however, neither the Ernst and Merchant nor the Lee and Shaffer theory approached quantitative agreement with any of the experimental relationships for the various materials tested. In fact, no unique relationship of the kind predicted by these theories could possibly agree with all the experimental

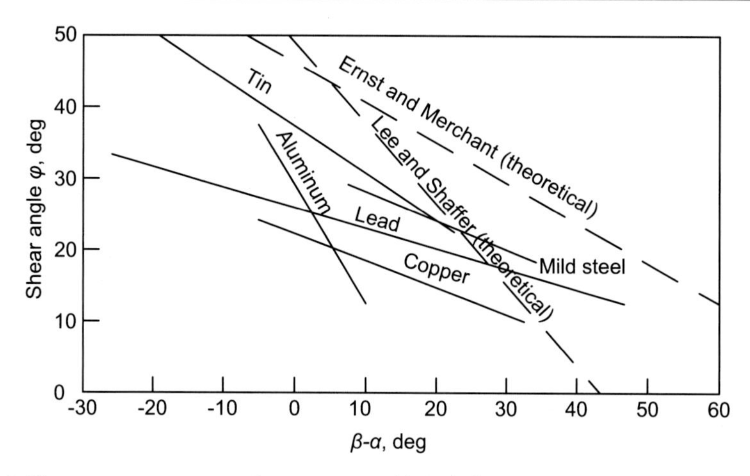

Fig. 7.9 Shear stress versus normal stress assumed in Merchant's second theory

results. Even the modified Merchant theory in which the shear stress on the shear plane is assumed to be linearly dependent on the normal stress could not agree with all the results. The modified Merchant theory yielded the relationship

$$2\varphi + \beta - \alpha = C, \tag{7.24}$$

where C is a constant depending on the work material. Substituting various values of C in the above could only give a family of parallel lines in Fig. 7.9. Evidently, the experimental lines are not parallel and could not be represented by Eq. (7.24) to a great degree of accuracy. The assumption that the cutting tool is perfectly sharp can contribute to the deviation between the theoretical and experimental shear angle values. This assumption is particularly rough at small undeformed chip thickness. In addition, the theories argued that the primary deformation zone can be regarded as a shear plane; however, a shear "zone" with finite thickness has been observed in many cutting experiments which suggested that the effects of strain hardening and strain rate effects ought not to be ignored as in the theories.

7.5 Velocities

There are three basic velocities involved in the orthogonal machining process. They are: (1) the cutting velocity (v), which is the velocity of the tool relative to the work and directed parallel to F_P in case of a relatively small feed, (2) the chip velocity (v_c), which is the velocity of the chip relative to the tool directed along the tool rake since the newly formed chip just rubs against the tool rake but never separates from the rake, and (3) the shear velocity (v_s), which is the velocity of the chip relative to the workpiece and directed along the shear plane since the newly formed chip just

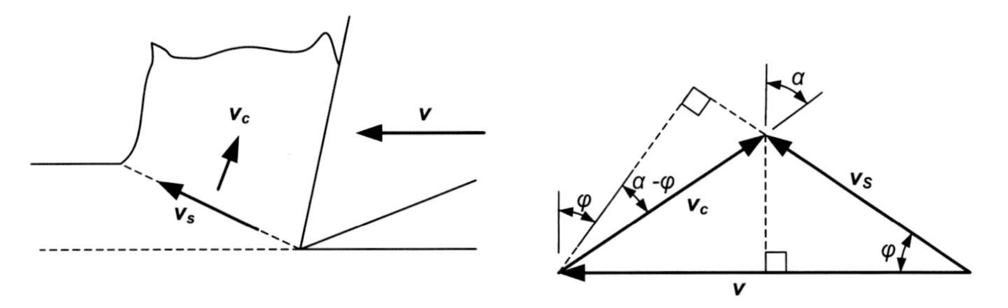

Fig. 7.10 Three basic velocities in orthogonal machining

slides through the shear plane but never separates from the plane. These velocities are shown in Fig. 7.10.

If the chip is incompressible, and the density is unaffected by the deformation process, based on continuity, the velocities are related by

$$v_s = v + v_c. \tag{7.25}$$

From the hodograph in Fig. 7.10, the following expressions can be derived:

$$v_c = v \frac{\sin \varphi}{\cos(\varphi - \alpha)} = vr, \tag{7.26}$$

$$v_s = v_c \frac{\cos \alpha}{\sin \varphi}, \tag{7.27}$$

and thus

$$v_s = v \frac{\cos \alpha}{\cos(\varphi - \alpha)}. \tag{7.28}$$

Example

The chip velocity is observed to be 13.2 in./s at a depth of cut of 0.001 in. and a spindle speed of 4 rps in the turning of a 1.5 in. diameter workpiece. The lathe bit has a rake angle of 5° and relief angle of 2°. Estimate the chip velocity if a lathe bit of 15° rake angle is used instead, and everything else remains unchanged.

Solution:

From Eq. (7.16), $r = \frac{v_c}{v} = \frac{13.2}{\pi(1.5)(4)} = 0.7$ and $\varphi = \tan^{-1} \frac{r \cos \alpha}{1 - r \sin \alpha} = 36.6^{\circ}$.

According to the Ernst and Merchant model: $\frac{\beta}{2} - \frac{\pi}{4} = \frac{\alpha}{2} - \varphi = \frac{\alpha'}{2} - \varphi'$

(This is assuming that the friction coefficient does not change with the rake angle)

$$\alpha = 5^{\circ}, \varphi = 36.6^{\circ}, \alpha' = 15^{\circ}, \text{ we have } \varphi' = 41.6^{\circ}$$

$$v'_c = r'v = \frac{\sin\varphi'}{\cos(\varphi' - \alpha')}\pi(1.5)(4) = 14 \text{ in./s.}$$

It appears that as the rake angle increases, the chip moves faster.

7.6 Shear Strain and Shear Strain Rate

To estimate the shear strain γ, triangle ABC in Fig. 7.11 is considered. It can be seen that

$$\gamma = \frac{\Delta S}{\Delta Y} = \frac{AB'}{CD} = \frac{AD}{CD} + \frac{DB'}{CD} = \tan(\varphi - \alpha) + \cot\varphi = \frac{\cos\alpha}{\sin\varphi\cos(\varphi - \alpha)}. \quad (7.29)$$

From Eqs. (7.28) and (7.29),

$$\gamma = \frac{v_s}{v\sin\varphi}. \quad (7.30)$$

The rate of strain is

$$\dot{\gamma} = \frac{\Delta S}{\Delta Y \Delta t} = \frac{v_s}{\Delta Y}. \quad (7.31)$$

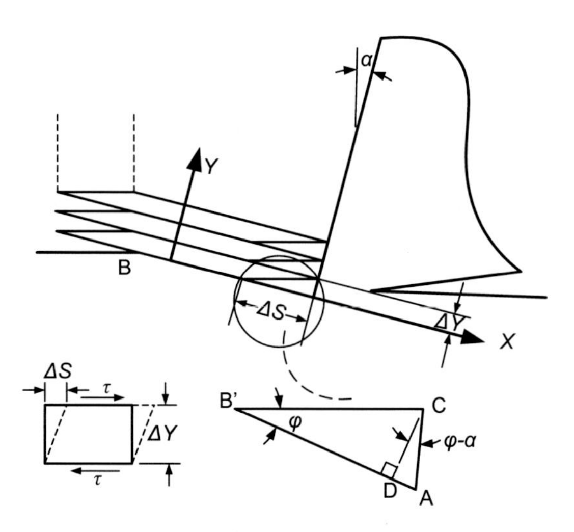

Fig. 7.11 Shear strain relationships

Photomicrographs of the shear plane show that the shear plane has very limited thickness, usually on the order of 10^{-3} in. For a shear velocity of 100 in./s, it implies that a strain rate can be on the order of 10^5 s^{-1}, which is 7~8 orders of magnitude higher than that in a typical uniaxial tensile test.

7.7 Cutting Energy

The total energy consumption rate in orthogonal machining is

$$U = F_P v. \tag{7.32}$$

The specific cutting energy, which is the total cutting energy required to remove unit volume of material, will therefore be

$$u = \frac{U}{vbt} = \frac{F_P}{bt}. \tag{7.33}$$

This specific cutting energy can be attributed to the following sources of energy dissipation: specific shear energy (u_S) on the shear plane, specific friction energy (u_F) on the tool face, specific surface energy (u_A) due to the formation of new surface areas in cutting, and as specific momentum energy (u_M) due to the momentum change associated with the material as it crosses the shear plane. That is

$$u = u_S + u_F + u_A + u_M. \tag{7.34}$$

The specific shear energy is the energy resisting plastic deformation along the shear plane. It can be given as

$$u_S = \frac{F_S v_s}{vbt} = \frac{v_s}{v} \frac{\tau}{\sin\varphi} \text{ (from Eq. (7.30))} = \tau\gamma. \tag{7.35}$$

The specific friction energy is

$$u_F = \frac{F_C v_c}{vbt} = \frac{F_C r}{bt}. \tag{7.36}$$

The specific surface energy is a term generally familiar to thermodynamics. It is understood to be the energy required to break the ion bonds at the generation of new surfaces, and it is usually related to the fracture toughness of the material. It can be described as

$$u_A = 2\frac{Tvb}{vbt} = 2\frac{T}{t}, \tag{7.37}$$

where T is the surface energy per unit area of the new surface created ≈ 0.006 in-lb/in^2 for metals. The factor of 2 accounts for the two surfaces generated.

The uncut portion of the chip is stationary with respect to the workpiece at the beginning, and it gains the velocity of the chip during cutting. Therefore, a certain amount of energy is consumed to change the momentum of the material as it crosses the shear plane. This energy per unit volume of material removed is the specific momentum energy given as

$$u_M = \frac{F_M v_s}{vbt}, \tag{7.38}$$

where F_M is the momentum force along the shear plane, and it is

$$F_M = \rho(vbt)v_s, \tag{7.39}$$

where ρ is the mass density of the workpiece. Therefore,

$$u_M = \rho v^2 \gamma^2 \sin^2 \varphi. \tag{7.40}$$

To gain some idea of the relative magnitude between the four specific energies, it is a good exercise to use the representative metal cutting data in Table 3.1 of Shaw [1984] and calculate the specific energies. For example, in the case of an undeformed chip thickness of 3.7×10^{-3} in. and a cutting speed of 1186 ft/min is considered in the fourth row of the Table, with $F_P = 303\,\text{lb}, F_Q = 168\,\text{lb}, b = 0.25\ \text{in}, \tau = 93 \times 10^3$ psi $\rho = 7.25 \times 10^{-4}\ \text{lb/in}^3, v = 1186\,\text{ft/min}, \gamma = 2.4, \varphi = 25^\circ, r = 0.44, \alpha = 10^\circ$, the following specific energy can be calculated:

$$u = \frac{F_P}{bt} = \frac{303}{(0.25)(3.7 \times 10^{-3})} = 327{,}567 \text{ psi}$$

$$u_S = \tau\gamma = (93 \times 10^3)(2.4) = 223{,}200 \text{ psi}$$

$$u_F = \frac{(F_P \sin\alpha + F_Q \cos\alpha)r}{bt} = \frac{(303 \sin 10^o + 168 \cos 10^o)0.44}{0.25(3.7 \times 10^{-3})} = 103{,}727 \text{ psi}$$

$$u_A = \frac{2T}{t} = \frac{2(0.006)}{3.7 \times 10^{-3}} = 3.2 \text{ psi}$$

$$u_M = \rho v^2 \gamma^2 \sin^2 \varphi = (7.25 \times 10^{-4})\left(\frac{1186 \times 12}{60}\right)^2 (2.4)^2 (\sin 25°) = 41.96 \text{ psi}$$

It can be seen in this example that most of the cutting energy is consumed in plastic deformation and friction. Momentum and surface energy is practically negligible.

Example
In a turning process, a tool with 8° rake angle cuts a cylindrical workpiece diameter from 2.8 in. down to 2.6 in. in one pass. With 600 rpm of spindle speed, the tool cuts through a 4 in. linear distance in 40 s while the tangential cutting force is measured to be 250 lb and thrust force 45 lb. Estimate the specific cutting energy, specific shear energy, and specific friction energy in the process.

Solution:

From Eq. (7.6), $\mu = \dfrac{F_Q + F_P \tan\alpha}{F_P - F_Q \tan\alpha} = \dfrac{45 + 250\tan 8^\circ}{250 - 45\tan 8^\circ} = 0.33.$

From Eq. (7.16), $\varphi = \dfrac{90^\circ - \tan^{-1} 0.33 + 8^\circ}{2} = 40^\circ.$

From Eq. (7.12), $r = \dfrac{\sin\varphi}{\cos(\varphi - \alpha)} = \dfrac{\sin 40^\circ}{\cos(40^0 - 8^\circ)} = 0.76.$

From Eq. (7.29), $\gamma = \tan(\varphi - \alpha) + \cot\varphi = \tan(40^\circ - 8^\circ) + \cot 40^\circ = 1.82.$

$$\text{uncut chip thickness } (t) = \frac{v}{\omega} = \frac{4/40}{600/60} = 0.01 \text{ in.}$$

From Eq. (7.9), $\tau = \dfrac{(F_P\cos\varphi - F_Q\sin\varphi)\sin\varphi}{bt} = \dfrac{(250\cos 40^\circ - 45\sin 40^\circ)\sin 40^\circ}{(0.1)(0.01)}$

$= 104{,}500$ psi.

From Eq. (7.33), $u = \dfrac{F_P}{bt} = \dfrac{250}{(0.1)(0.01)} = 250{,}000$ psi.

From Eq. (7.35), $u_S = \tau\gamma = (104{,}500)(1.82) = 190{,}190$ psi.

From Eqs. (7.36) and (7.4),

$$u_F = \frac{(F_P\sin\alpha + F_Q\cos\alpha)r}{bt} = \frac{(250\sin 8^\circ + 45\cos 8^\circ)(0.76)}{(0.1)(0.01)} = 60{,}148 \text{ psi.}$$

Although the specific cutting energy, u, is dependent on material properties (especially the hardness), it can vary considerably for a given material and is affected by changes in tool rake and feed.

A larger rake will require less amount of plastic deformation of the chip and therefore a smaller specific cutting energy. It is commonly experienced that every degree increase of effective rake angle results in 1 % decrease in u.

No cutting tool is perfectly sharp, and cutting edge is usually represented by a cylindrical surface, i.e., a tool nose, joining the tool flank and the tool face. As the tool edge "plows" its way through the work material, there exists a force acting over the tool nose area. This force, usually called the "plowing force," forms only a small proportion of the cutting force at large undeformed chip thickness. At small undeformed chip thickness, however, the plowing force is proportionately large and cannot be neglected. When the total cutting force is divided by the undeformed-chip cross-sectional area to give u, the portion of u contributing to chip removal will remain constant, and the portion resulting from the plowing force will increase as the chip thickness decreases. As a rule of thumb, u is proportional to the (-0.2)th order of the undeformed chip thickness. The increase in specific cutting energy as the chip thickness decreases explains why processes, such as grinding, that produce very thin chips require greater power to remove a given volume of material.

Homework

1. In a turning process a lathe bit of 90° major cutting-edge angle, 15° rake angle, and 5° clearance angle is used at a feed of 0.005 ipr, radial depth of cut of 0.1 in., and spindle speed of 1200 rpm. The apparent shear strength of the workpiece material is 127 kpsi. Knowing the friction coefficient between the tool rake face and the chip is about 0.454, estimate the resultant cutting force in the process.
2. In the single-point turning of cast iron with a high-speed steel (HSS) tool, the tangential cutting force and the thrust force are measured to be 250 and 48 lb, respectively, at a radial depth of cut of 0.05 in. and a spindle speed of 1200 rpm. The workpiece has a diameter of 2.5 in. and a total length of cut of 6 in. to be completed in 1 min time. Estimate the apparent shear strength of the workpiece material. Note that the major cutting-edge angle of the tool is 90°, and the process configuration can be considered as orthogonal machining.
3. An orthogonal machining experiment was run using a cutting edge with a rake angle of 7° and flank angle of 5°. The uncut chip thickness was 0.5 mm. The measured chip thickness was 0.9 mm under dry cutting condition, while it was 0.7 mm in wet cutting. Estimate the friction coefficient based on the Ernst and Merchant model and the Lee and Shaffer model.
4. Suppose a 50 mm diameter annealed 1113 steel bar with a shear strength of 210 MPa has to be plunge turned by a tool with a 10° rake angle and 90° major cutting-edge angle. The width of cut is 5 mm, depth of cut (uncut chip thickness) is 1 mm, and spindle speed is 5200 rpm. Estimate the amount of power required to perform this operation, the specific cutting energy of the material under this condition, and the strain rate on the shear plane. Assume that the friction coefficient is 0.3, and apply the Ernst and Merchant model.
5. In the development of the shear angle solution by Ernst and Merchant, it was assumed that the shear stress remains constant with various shear angles φ. This was the basic condition that led to Eq. (7.16) by differentiating Eq. (7.15) with respect to φ. Suppose you now propose a new assumption stating that the normal stress, σ, instead of the shear stress, is invariant to the shear angle. What should be the shear angle solution as a result of this assumption?

Shear Stress in Cutting

8

A more careful examination of the shear stress that occurred on the shear plane in metal cutting is made in this chapter. It is motivated by a need to explain the significant difference between the shear-stress and shear-strain values in orthogonal cutting data and torsional tests data. If the following equations derived in Chap. 7 are used and plotted as shown in Fig. 8.1, it can be observed that typical cutting data appear much higher than torsional data.

$$\tau = \frac{\left(F_P \cos\varphi - F_Q \sin\varphi\right)\sin\varphi}{bt}, \tag{8.1}$$

$$\gamma = \tan(\varphi - \alpha) + \cos\varphi. \tag{8.2}$$

The implication of this difference is that the results from standardized tests (torsional or tensile) cannot be adopted to estimate the level of shear stresses and shear strains, as well as cutting forces and power, experienced in metal cutting.

Plastic deformation of workpiece on the shear plane in orthogonal cutting with continuous chip formation may be characterized by the large values of strain rate and temperature and the existence of normal stress on the shear plane resulting in large values of homogeneous strain without gross fracture. These conditions can be very different than those in standardized tests and their effects on the shear stress on the shear plane will be discussed herein.

8.1 Strain Rate and Temperature

It has been suggested that the thermal energy of a material is available to assist the applied stress in forming dislocations on the shear plane in cutting. Figure 8.2 shows an elastic stress–strain curve for a perfectly homogeneous material. The energy required to raise a unit volume of material to the flow stress level τ_o is represented by the shaded area. The stress τ is the apparent strength defined by the

S. Y. Liang, A. J. Shih, *Analysis of Machining and Machine Tools*,
DOI 10.1007/978-1-4899-7645-1_8

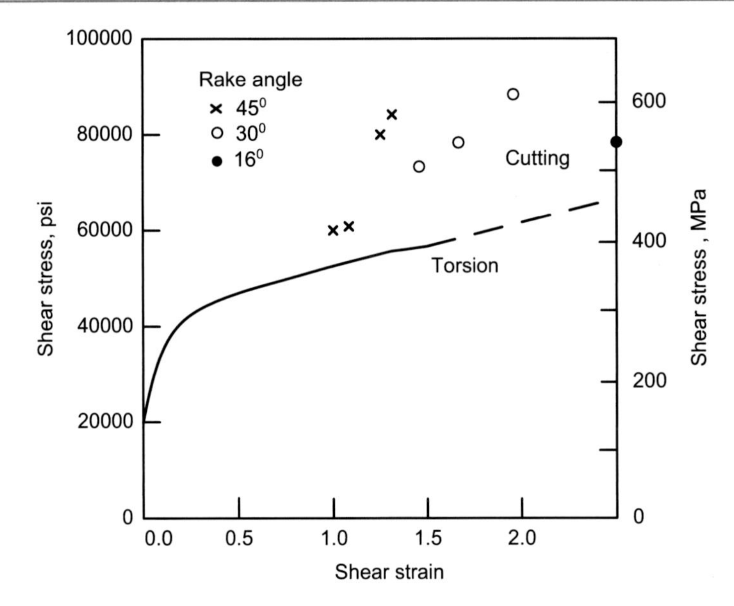

Fig. 8.1 Discrepancy between cutting and torsion data of AISI B1112 steel cut at a low cutting speed

Fig. 8.2 Stress–strain curve

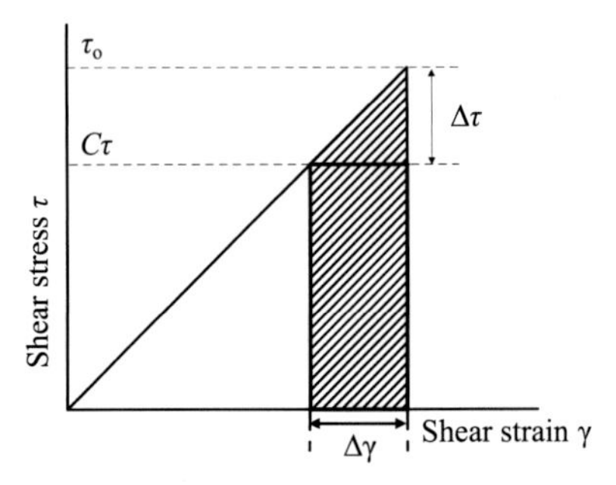

applied stress from an external force field and C is the stress concentration factor at the tip of microcracks on the shear plane. The energy required to shear the work material in this case is the entire area given by $\frac{\tau_o^2}{2G}$, where G is the shear modulus and $\tau_o = C\tau + \Delta\tau$. The portion of this required energy attributed to the assistance from thermal energy u_θ is thus

$$u_\theta = \frac{\Delta\tau^2}{2G} + C\tau\Delta\gamma = \frac{\Delta\tau^2}{2G} + \frac{C\tau\Delta\tau}{G} = \frac{\left(\tau_o^2 - C^2\tau^2\right)}{2G}. \tag{8.3}$$

This suggests that a dislocation will be formed whenever the thermal energy per unit volume in the vicinity of a stress concentration reaches the value specified by the above equation.

Boltzmann (1956) has shown that the probability P, of finding a thermal energy $u_\theta V'$, in a volume of V' is

$$P = \exp\left(\frac{-u_\theta V'}{kT}\right), \tag{8.4}$$

where T is the absolute temperature and k is the Boltzmann's constant with a value of 1.38×10^{-16} erg/K.

The rate of plastic strain $\dot{\gamma}$ is given by

$$\dot{\gamma} = b\rho\bar{v}, \tag{8.5}$$

where b is the Burgers vector with a magnitude equivalent to one atom spacing of the material, ρ is the density of dislocation, and $\bar{v}$ is the average velocity of the dislocations. It is believed that the thermal energy probability P is proportional to the density of dislocation, and the rate of plastic strain is also proportional to the probability of there being sufficient thermal energy to form a dislocation. Therefore,

$$\gamma = \mathrm{AP} = A\exp\left(-\frac{u_\theta V'}{kT}\right), \tag{8.6}$$

where A is a proportionality constant. With the value of u_θ given by Eq. (8.3), the equation above becomes

$$\dot{\gamma} = A\exp\left(\frac{-\left(\tau_o^2 - C^2\tau^2\right)V'}{2GkT}\right), \tag{8.7}$$

which can be arranged into

$$\tau^2 = \frac{\tau_o^2}{c^2} - T\left[\frac{2Gk}{c^2V'}(\ln A) - \frac{2Gk}{c^2V'}(\ln\dot{\gamma})\right]. \tag{8.8}$$

For a given material τ_o, c, G, k, V' and A are all constants; therefore, we may define

$$C_1 = \frac{\tau_o^2}{c^2}, C_2 = \frac{2GK}{c^2V'}(\ln A), \text{ and } C_3 = \frac{2Gk}{c^2V'}, \tag{8.9}$$

where C_1, C_2, and C_3 are property-dependent constants. Equation (8.8) can thus be expressed as

$$\tau^2 = C_1 - T\left[C_2 - C_3(\ln\dot{\gamma})\right]. \tag{8.10}$$

The tensile data for the shear stress of annealed AISI B1112 steel at different temperatures and strain rates is as shown in the table below.

Absolute temperature (T), R	Strain rate ($\dot{\gamma}$), s^{-1}	Shear stress (τ), psi
530	0.0009	38,550
530	482	46,400
160	0.0009	59,700

With the above data, constants C_1, C_2, and C_3 can be solved for, and the expression of Eq. (8.10) becomes

$$\tau = \left(4.4\times10^9 - T\left[4.9\times10^6 - 9.5\times10^3(\ln\dot{\gamma})\right]\right)^{1/2}. \tag{8.11}$$

It concludes that the effect of temperature on the apparent shear strength of the material is offset by the effect of strain rate, and vice versa.

8.2 Effect of Normal Stress

The shear stress on shear plane in cutting is also affected by the existence of normal stress. Bridgman in 1952 conducted torsional tests on the hollow tubular notched specimen shown in Fig. 8.3. The specimen was also subjected to axial compression in addition to torsional shear. Due to the reduced cross section at the notch, strain was concentrated in these notch sections. Shear strain versus shear stress curves were estimated under various loading conditions and it was found that the flow curve was the same for all values of compressive stress on the shear plane, a result consistent with other materials tests involving much lower plastic strains. However, the strain at gross fracture was found to be strongly influenced by the existence of compressive stress, increasing markedly with the increase in compressive stress.

Walker and Shaw, in 1967, measured the acoustical energy resulting from plastic flow during the torsional and compressive loading of tubular specimens. It was observed that a region of rather intense acoustical activity occurred at the yield point followed by a quieter regime until a shear strain of about 1.5 was reached. To explain this behavior, a mechanism of large strain plastic flow was subsequently developed in which it was suggested that at moderate values of normal stress on the shear plane, microcracks begin to appear in a plane of concentrated shear at a shear strain of about 1.5. As strain proceeds beyond this point, the first microcracks are sheared shut as new ones take their place. The quiet area on the shear plane gradually increases until it becomes insufficient to resist the shear load without fracture. This mechanism appeared as a deviation from the dislocation theory, but it provided

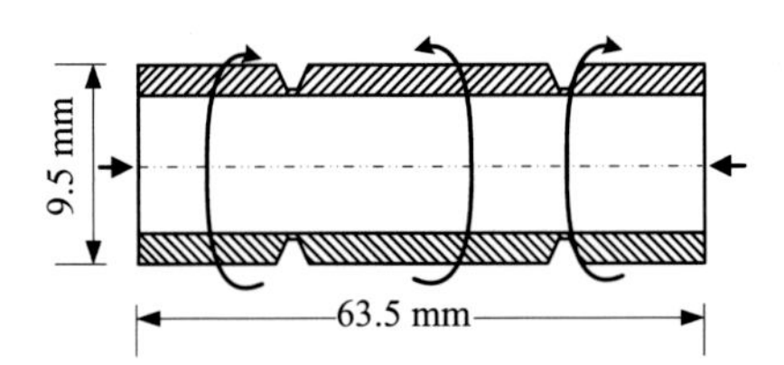

Fig. 8.3 Combined axial and torsional loads used in Bridgman specimen

an explanation for the infidelity of usual strain-hardening curves in describing the behavior of large plastic flow.

If the normal stress on the shear plane is tensile, these microcracks will spread rapidly over the shear surface leading to gross fracture. In metal cutting, the presence of compressive stress on the shear plane suppressed the prorogation of microcracks and increased the required shear stress for deformation. Ordinary material tests either do not involve microcrack formation or have a ratio of normal stress to shear stress on the shear plane that is quite different from that in cutting.

8.3 Inhomogeneous Strain

The commonly accepted stress–strain relationship in plasticity follows the strain-hardening curve as

$$\sigma = K\varepsilon^n, \tag{8.12}$$

in which K is referred to as the strength coefficient and n is the strain-hardening exponent. However, experimental proof from Blazynski and Cole in 1960 shows that the equation above is good only up to a strain of about 1. Beyond a strain of 1, the stress–strain relationship is linear and can be given by

$$\sigma = (1-n)K + nK\varepsilon \equiv A + B\varepsilon. \tag{8.13}$$

Although the equation above is proposed for normal stress, shear stress can also be approximated to be in a linear relationship with the extent of slip that occurs on a given shear plane,

$$\tau = A' + B'\Delta S. \tag{8.14}$$

In the shear deformation of crystalline metal materials, it is found that slip does not occur uniformly on every atomic plane but the active slip planes are relatively far apart. This blockwise deformation is believed to be attributed to the inherent imperfections in the crystal. As the volume of material deformed is relatively large, uniform density of imperfections can be assumed and the strain can be considered uniform. However, as the volume deformed approaches a small volume comparable to that of an imperfection, the blockwise inhomogeneous characteristic of strain becomes rather obvious. Therefore, there is a size effect on the strain inhomogeneity.

In metal cutting, the undeformed chip thickness is usually very small compared to the width of cut and the inhomogeneous strain can often be observed at the back of a chip as seen in Fig. 8.4. The shear plane in macroscopic view is actually the accumulation of a series of active slip planes each extended for a distance of a along the width of the chip. If active planes are introduced by crystalline imperfections, the spacing between these shear planes will correspond to the closeness of the imperfections.

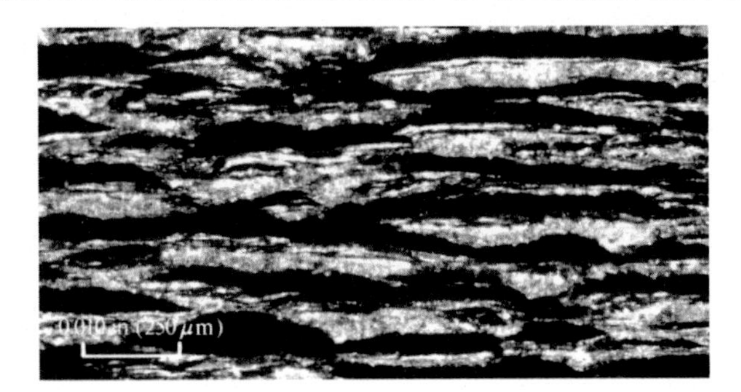

Fig. 8.4 Concentrated strain on the back of chip

To facilitate analysis, it can be assumed that the workpiece material is filled with an orderly array of imperfections with a uniform spacing of a in each direction as seen in Fig. 8.5. Let P_1 and P_2 be two parallel shear planes making an angle φ with the direction of cut and passing through adjacent points in the first row below the surface. For a depth of cut t, there will be (t/a) active shear planes placed between P_1 and P_2 such that a single plane passes through each imperfection in the layer in the process of cutting. Note that these t/a active shear planes do not have to be parallel to one another. Since the distance between P_1 and P_2 is $a \sin \varphi$, the average spacing between active shear planes is

$$\Delta Y = \frac{a \sin \varphi}{t/a} = \frac{a^2 \sin \varphi}{t} \tag{8.15}$$

and from Eq. (7.29) the amount of slip on an active shear plane is

$$\Delta S = \frac{a^2 \sin \varphi}{t} \gamma. \tag{8.16}$$

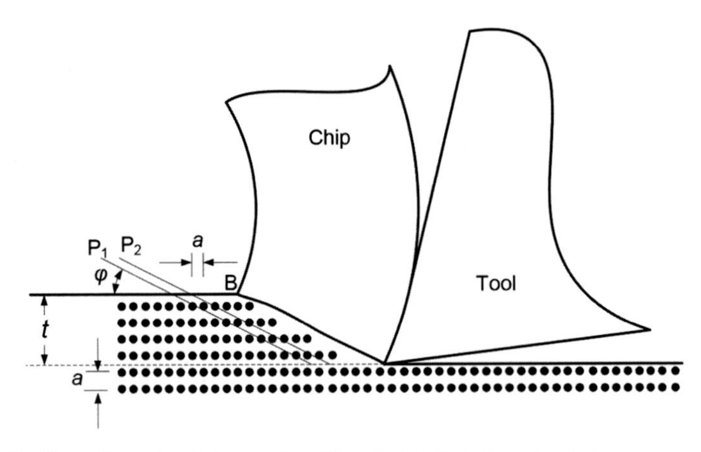

Fig. 8.5 Configuration of workpiece with uniformly distributed weak points

The shear stress on the macroscopic shear plane is actually equivalent to the stress on each active microscopic shear plane. From Eqs. (8.14) and (8.16) above and Eq. (7.29), it can be shown that

$$\tau = A' + B' \frac{a^2 \sin\varphi}{t}\left[\cot\varphi + \tan(\varphi - \alpha)\right]. \tag{8.17}$$

This equation suggests that the shear stress on the shear plane is not constant but depends upon the rate of strain hardening (as influenced by A' and B'), the strain in the chip (ΔS or γ), the density of imperfections a, and the undeformed chip thickness relative to the imperfection density (t/a).

Example

The inhomogeneous strain model as expressed by Eq. (8.17) can be evaluated with the represented data in Table 3.1 in Shaw (1984). Suppose the following cutting data are available from machining NE9445 steel having a Brinell hardness of 187 with sintered carbide tool. The width of cut was 0.25 in.

t (inch)	φ	a	τ (psi)
3.7×10^{-3}	17°	10°	85×10^3
1.09×10^{-3}	19°	10°	103×10^3

Estimate the shear stress of material on the shear plane for a depth of cut of 2.34×10^{-3} in., shear angle of 18.5°, and rake angle of 10°.

Solution

Substituting the above data into Eq. (8.17) yields $A' = 77{,}440$ psi and $B' a_2 = 28.5$ lb/in. Assuming the imperfection density does not change between different work batches, Eq. (8.17) can be used again with $t = 2.34 \times 10^{-3}$, $\varphi = 18.5°$, and $a = 10°$ to give $\tau = 90 \times 10^3$ psi. Note that in Table 3.1 (Shaw 1984), this cutting condition has $\tau = 93 \times 10^3$ psi.

9 Cutting Temperature and Thermal Analysis

During cutting, high temperatures are generated in the region of the tool cutting edge as a form of cutting energy dissipation. These temperatures have a controlling influence on the rate of wear of the cutting tool and on the friction between the chip and tool, and they can significantly affect the functional performance of a machined part due to residual stress or thermal distortion. Therefore, considerable attention has been paid to the measurement and prediction of the temperatures in the tool, chip, and workpiece in metal cutting.

For example, experiments have shown that when copper is turned with a carbide tool under 0.3 mm radial depth of cut, 0.2 mm/rev feed, and 100 m/min feed rate, the cutting temperature is around 300 °C. For a copper workpiece with a diameter of 30 mm and a linear coefficient of thermal expansion (CTE) at 20 °C of 17×10^{-6}/°C, the diameter variation due to thermal effect would be $30 \times 17 \times 10^{-6} \times 300 = 0.153$ mm. Therefore, the size of workpiece measured after it is cooled would be smaller than it is during machining by about 50 % of the depth of cut. Of course, this estimate involves many unjustified assumptions including the uniformity of cutting temperature distribution and the constant CTE. However, it gives us a rough sense of the possible effect of cutting temperature on the process performance.

9.1 The Measurement of Cutting Temperatures

A number of methods have been developed for the temperature measurements in metal cutting. Some of these only measure the average cutting temperature, although methods for determining temperature distribution in the cutting zone under carefully controlled laboratory conditions are possible.

S. Y. Liang, A. J. Shih, *Analysis of Machining and Machine Tools*,
DOI 10.1007/978-1-4899-7645-1_9

9.1.1 Thermocouple Method

The method most widely used to determine cutting temperature is the thermocouple technique. It can be applied either in a work-tool thermocouple configuration or embedded thermocouple configuration. Figure 9.1 shows a work-tool thermocouple arrangement on a lathe. The thermocouple circuit is insulated from the machine, and the same circuit is used to calibrate the thermocouple. This arrangement is limited to the measurement of mean temperature along the chip–tool interface without indication of the distribution of temperature. Figure 9.2 shows the setup of an embedded thermocouple and typical temperature–time traces. It offers the temperature reading

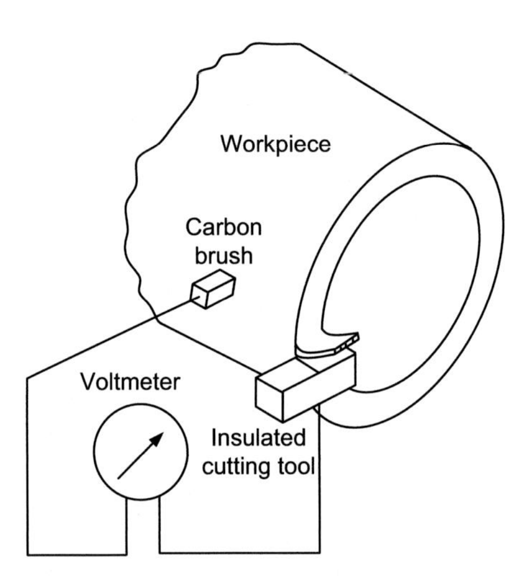

Fig. 9.1 Work-tool thermocouple device for cutting temperature measurement

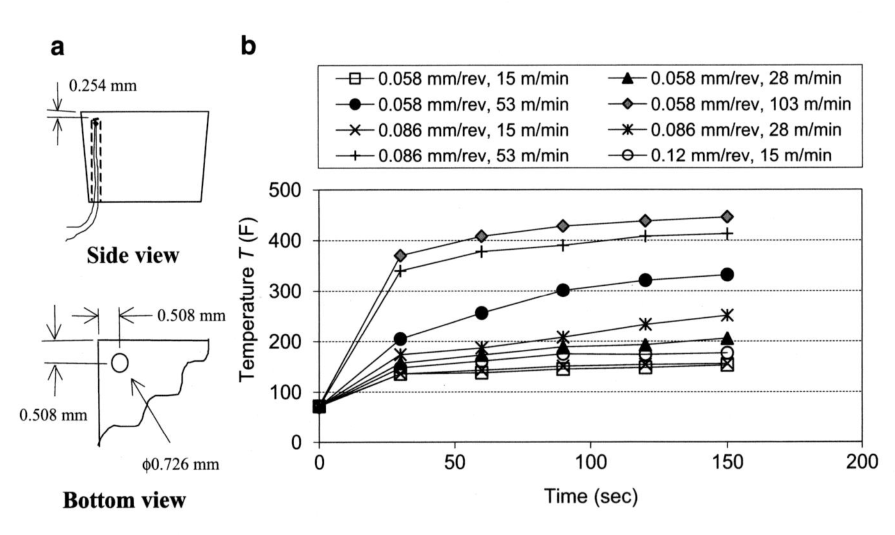

Fig. 9.2 **a** Typical embedded thermocouple set-up and **b** tool tip temperature in turning of Al6061-T6 with a carbide insert

at a localized point close to the cutting tool surface. Both work-tool and embedded methods have been used extensively to investigate the effects of changes in cutting conditions on cutting temperatures and to obtain empirical relationships between temperature and cutting tool wear rate.

9.1.2 Radiation Method

Cameras and films sensitive to infrared radiation can be used to determine the temperature distribution in the cutting zone. A furnace of known temperature is usually photographed simultaneously with the cutting operation to allow calibration. Although the current improvements in infrared-sensitive films and developments of thermal imaging video cameras now make it possible to determine temperature of workpiece from room temperature to well over 1000 °C, the technique is applicable only to cutting processes with visually observable tool-workpiece area. This factor limits the technique to no more than laboratory testing purposes. Figure 9.3 shows a typical digital image obtained with an infrared camera during a tube-facing operation.

9.1.3 Hardness and Microstructure Method

The room-temperature hardness of hardened steel decreases after reheating and the loss in hardness is related to the temperature and time of heating. The hardness decrease is the result of changes in the microstructure of the steel. These structural changes can be observed using optical and electron microscopes. These changes provide an effective means of determining temperature distributions in the tool during cutting. Microhardness measurements on tools after cutting can be used to determine constant-temperature contours in the tool, but the technique is time-consuming, requiring very accurate hardness measurements, and relying upon experienced interpretation of the observed structural changes. For steels, the commonly achievable accuracy is on the order of ± 25 °C at the current state of the art.

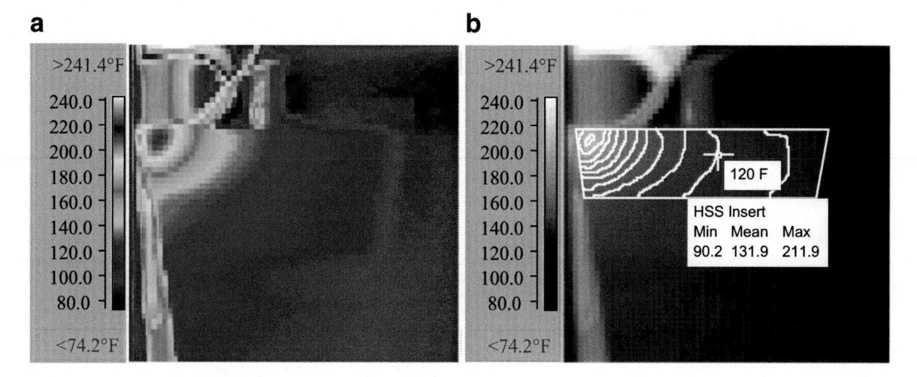

Fig. 9.3 **a** Infrared thermal image and **b** processed image with 10 °F isotherms for dry cut of Al6061-T6 with HSS at 28 m/min, 0.058 mm/rev

9.2 Heat Generation in Machining

When a material is deformed elastically, the energy required is stored in the material as strain energy, and no heat is generated. However, when a material is deformed plastically, most of the energy used is converted into heat. In metal cutting, the material is subject to extremely high strains. Since the elastic deformation forms only a very small portion of the total deformation, it may be assumed that all the energy required in cutting is converted into heat. It is recalled that the rate of energy consumption during machining, P, is given by

$$P = F_P v, \tag{9.1}$$

where F_P is the cutting component of the resultant tool force and v is the cutting speed.

In cutting, there are two major regions of plastic deformation responsible for the conversion of energy into heat: shearor primary deformation zone and secondary deformation zone as seen in Fig. 9.4. If the tool is not severely worn, the workpiece deformation at the contact with the tool flank can be considered negligible. Thus,

$$P = P_S + P_F, \tag{9.2}$$

where P_S is the rate of heat generation in the primary deformation zone (shear-zone heat rate) and P_F is the rate of heat generation in the secondary deformation zone (frictional heat rate). P_F is given by $F_C v_c$, where F_C is the frictional force on the tool face and v_c is the velocity of chip flow, which is given by rv.

To understand how heat is removed from these zones by the workpiece, chip, and tool motions, it is necessary to consider the transfer of heat in a material moving relative to a heat source. Figure 9.5 shows four stationary points located in a Cartesian coordinate. There is a material flowing through these points in the X direction. The material at point A is assumed to have an instantaneous temperature T. Coordinates and temperatures at points A, B, C, and D are shown in the figure.

Heat is transferred across the boundaries AB and CD by the combination of conduction due to the temperature gradients in the X direction and by convection because of the flow of heated material across these boundaries. Across BC and AD,

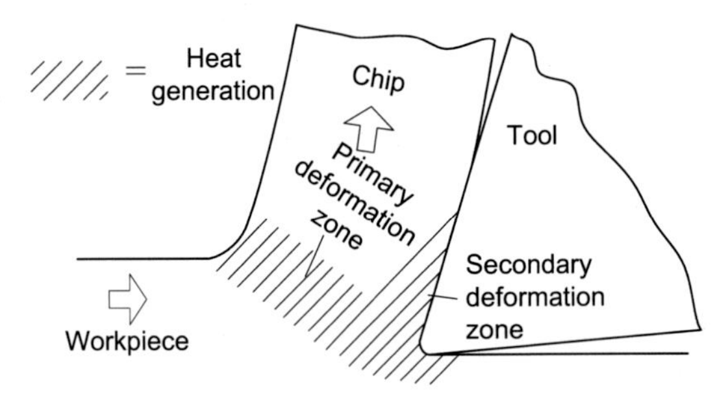

Fig. 9.4 Heat sources in orthogonal cutting

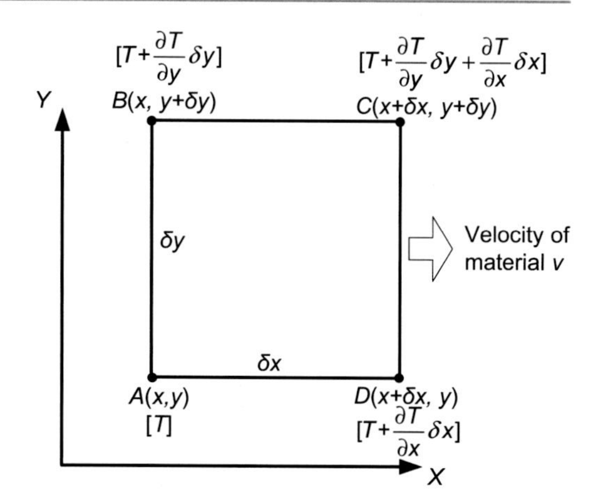

Fig. 9.5 Flow of heated material through an element *b*

heat can only be transferred by conduction since there is no material flowing across these boundaries. If there is a heat source with intensity of *P* (time rate of heat generation) within the element defined by ABCD, the balance of energy requires that

$$\frac{\partial^2 T}{\partial x^2}+\frac{\partial^2 T}{\partial y^2}+\left(\frac{R}{b}\frac{\partial T}{\partial \mathrm{x}}\right)=P, \tag{9.3}$$

where *R* is a dimensionless number given by $\rho cvb/k$ and is known as the thermal number with ρ being the density (kg/m^3), *k* the thermal conductivity (W/m-K), *c* the specific heat capacity (J/(kgK)), *v* the velocity of the material (m/s), and *b* the linear dimension along the lateral (Z) direction (width of cut). The first term in the equation is attributed to the conduction across AB and CD, the second term to the conduction across BC and AD, and the third term to the convection through AB and CD. Note that *b* should be equivalent to δy in Fig. 9.5. For single-point turning, *b* is typically the width of cut (in the radial direction).

The solution of Eq. (9.3) is possible with simple boundary conditions. Figure 9.6 shows its solution for a 1-D case, in which a point in the material approaching the heat source is heated rapidly, reaches its maximum temperature at the heat source, and then remains at constant temperature.

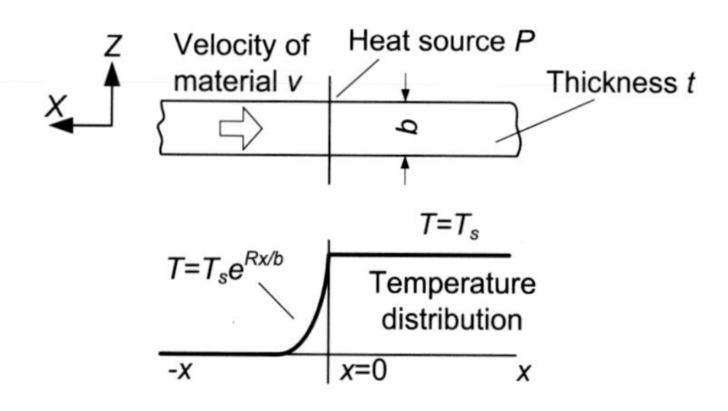

Fig. 9.6 Temperature distribution in a fast-moving material for a 1-D case, in which $T_s=P/\rho cvtb$. Note that the *P* is in the unit of W (J/s)

9.3 Temperature Distribution in Machining

Figure 9.7 shows the experimentally determined temperature distribution in the workpiece and the chip during orthogonal metal cutting. As a point X in the material, moving toward the cutting tool, approaches and passes through the primary deformation zone, it is heated until it leaves the zone. Point Y, however, passes through both deformation zones and is heated until it leaves the region of secondary deformation. Thus the maximum temperature occurs along the tool face with some distance from the cutting edge. Point Z, which remains in the workpiece, is heated by the conduction of heat from the primary deformation zone. Some heat is conducted from the secondary deformation into the body of the tool as well. In summary,

$$P = \Phi_c + \Phi_w + \Phi_t, \tag{9.4}$$

where P = total rate of heat generation, Φ_c = rate of heat transportation by the chip, Φ_w = rate of heat conduction into the workpiece, and Φ_t = rate of heat conduction into the tool. The chip material at the tool face usually flows rather fast, thus Φ_t forms a very small proportion of P and may be neglected except at a very low-cutting speed.

9.3.1 Temperatures in the Primary Deformation Zone

The rate of heat generation in the primary deformation (shear) zone is P_S. A fraction, Γ, of this heat is conducted into the workpiece, and the remainder is trans-

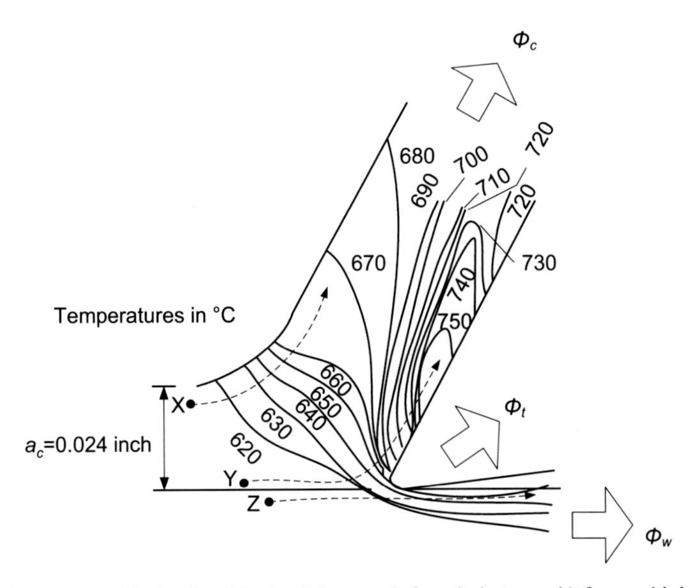

Fig. 9.7 Temperature distribution (obtained from an infrared photograph) for machining of free-cutting mild steel at a speed of 75 ft/min, width of cut of 0.25 in., by a tool rake of 30°. The workpiece temperature is 611 °C

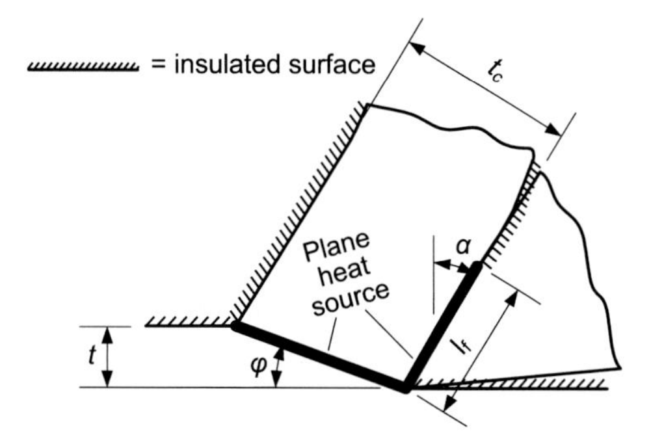

Fig. 9.8 Idealized cutting model in theoretical study on cutting temperature

ported with the chip. Thus, the average temperature rise, T_S, of the material passing through the primary deformation zone is given by

$$T_S = \frac{(1 - \Gamma)P_S}{\rho c v t b}, \tag{9.5}$$

in which t is the uncut chip thickness and b is the width of cut. Figure 9.8 shows an idealized model of cutting process employed by Rapier in 1954. This model assumed that the primary deformation zone could be regarded as a uniform plane heat source that no heat was lost from the free surfaces of the workpiece and chip, and that the thermal properties of the work material were constant and independent of temperature. To solve for Eq. (9.3), Weiner in 1955 further suggested that no heat was conducted into the material in the direction of its motion, since the transfer of heat in this direction of motion is mainly due to convection at high speeds. Equation (9.3) then simplifies to

$$\frac{\partial^2 T}{\partial y^2} + \frac{R}{b}\frac{\partial T}{\partial x} = P. \tag{9.6}$$

The solution to Eq. (9.6) with the stipulated boundary conditions as seen in Fig. 9.8 is provided in Fig. 9.9. Compared with experimental data, it is seen that the theory provides slightly underestimated results at high values of $R\tan\varphi$ (i.e., at high speeds and feeds). In the theory where a plane heat source was assumed, heat can only flow into the workpiece by conduction. In reality, heat is generated over a wide zone, part of which extends into the workpiece. The effect of this wide heat generation zone becomes increasingly important at high speeds and feeds, and therefore the deviation between theoretical and experimental data could be explained.

9.3.2 Temperatures in the Secondary Deformation Zone

The maximum temperature in the chip takes place at the exit from the secondary deformation zone, point C in Fig. 9.8. It is given by

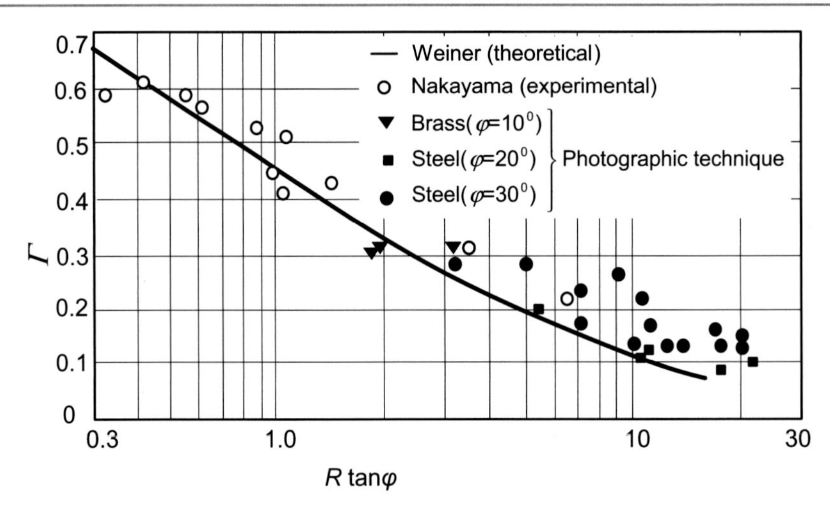

Fig. 9.9 Effect of $R\tan\varphi$ on division of shear-zone heat between chip and work, where Γ = the proportion of shear zone conducted into the work, R = thermal number and φ = shear angle

$$T_{\max} = T_o + T_s + T_m, \tag{9.7}$$

where T_o = initial workpiece temperature, T_s = temperature increase as material passes through the primary deformation zone, and T_m = temperature increase as material passes through the secondary deformation zone. Also, Fig. 9.8 shows the idealization by Rapier that the heat source resulting from the friction between the chip and the tool face was a plane heat source of uniform strength over the length of l_f. The following solution was obtained:

$$T_m = 1.13 T_f \sqrt{\frac{R}{l_o}}, \tag{9.8}$$

where l_o is the ratio of the heat source length to the chip thickness (l_f/t_c) and T_f is the average temperature increase of the chip, resulting from the secondary deformation as given by

$$T_f = \frac{P_F}{\rho c v t b}. \tag{9.9}$$

A comparison of Eq. (9.9) with experimental data by Rapier showed that his theory considerably overestimated T_m. This could be explained again by the fact that the friction-deformation zone, instead of being planar, has a finite width. The boundary conditions that more closely approximate reality is shown in Fig. 9.10, and an analysis based on this revised model by Boothroyd in 1963 yielded better agree-

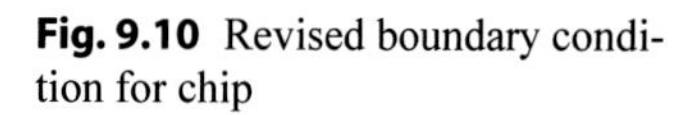

Fig. 9.10 Revised boundary condition for chip

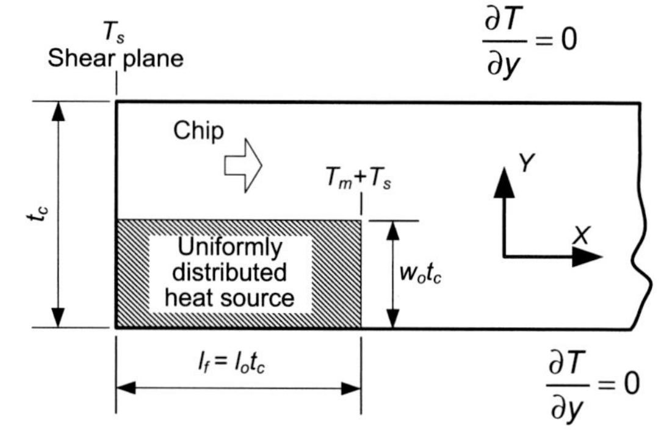

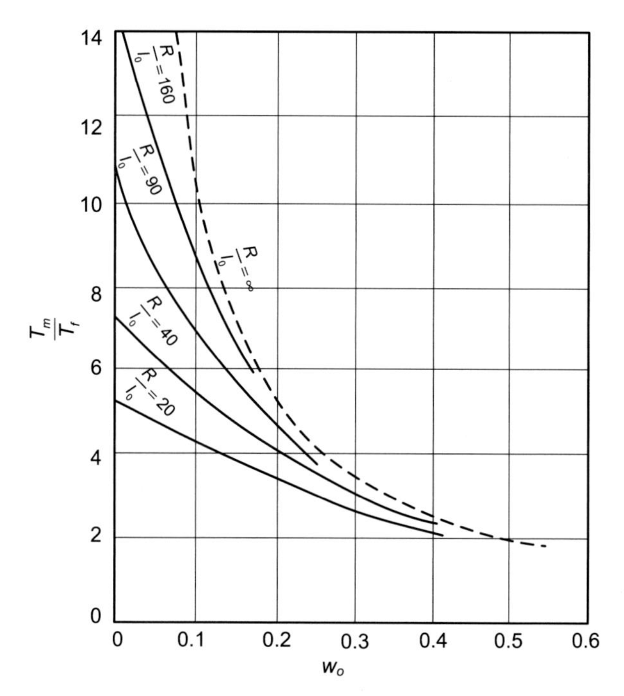

Fig. 9.11 Effect of secondary deformation zone width on chip temperatures, where R = thermal number, $l_o t_c$ = chip–tool contact length, $w_o t_c$ = width of secondary deformation zone, T_m = maximum temperature rise in chip, and T_f = mean temperature rise in chip

ment with experimental data. Figure 9.11 gives these results in terms of the effect of the zone width ($w_o t_c$) on temperature increase. When these curves are used, l_o can be estimated from the wear marks on the tool face, and the zone width can be estimated from a photomicrograph of the chip cross section as shown in Fig. 9.12.

Fig. 9.12 Grain deformation in chip cross section

Example

The maximum temperature along the tool face and the temperature on the newly machined surface are to be estimated for the following condition during the turning of mild steel at a room temperature of 22 °C:

Rake angle $\alpha=0$
Cutting force $F_P=890$ N
Thrust force $F_Q=667$ N
Cutting speed $v=2$ m/s
Undeformed chip thickness $t=0.25$ mm = feed
Width of cut $b=2.5$ mm
Cutting ratio $r=0.3$
Length of contact between chip and tool face $l_f=0.75$ mm
Width of secondary deformation zone under unlubricated condition $w_o=0.2$
Mild steel density $\rho=7200$ kg/m^3
Mild steel thermal conductivity $k=436$ W/m-K
Mild steel specific heat capacity $c=502$ J/kgK

Solution:

The total heat generation rate is $P = F_P v = 890(2) = 1780$ J/s.

In this example, $\alpha=0$, therefore $F_C = -F_P \sin\alpha - F_Q \cos\alpha = -F_Q$. Thus, the rate of heat generated by friction between the chip and the tool is given by

$$P_F = F_C v_c = F_C vr = 667(2)(0.3) = 400 \text{ J/s}.$$

From Eq. (9.2), the rate of heat generation from shearing in the primary deformation zone is $P_S = P - P_F = 1380$ J/s.

To estimate the temperature increase T_s, Fig. 9.9 is first used to find Γ for a given $R\tan\varphi$ value. The thermal number is $R = \frac{7200(502)(2)(0.0025)}{436} = 41.5$.

As the rake angle equals 0, $\tan\varphi = r\cos\alpha/(1-r\sin\alpha)$, thus

$$R\tan\varphi = 41.5(0.3) = 12.45.$$

From Fig. 9.9, Γ is found to be 0.1, and it implies that the majority of the heat is transported with the chip instead of being conducted into the workpiece. The temperature increase through the primary deformation zone is

$$T_S = \frac{(1-\Gamma)P_S}{\rho cvtb} = \frac{(1-0.1)1380}{7200(502)2(0.00025)0.0025} = \frac{(1-0.1)1380}{4.518} = 275°\text{C}.$$

To have a reasonable estimate of the temperature increase across the friction deformation zone, Fig. 9.11 instead of Eq. (9.8) is used here. Noting that $w_o = 0.2$ and

$$\frac{R}{l_o} = \frac{R}{l_f(r/t)} = \frac{(41.5)}{0.75(0.3/0.25)} = 46.1,$$

the ratio T_m/T_f can be looked up from the figure to be 4.2 and

$$T_f = \frac{400}{4.518} = 88.5°\text{C},$$

therefore $T_m = 4.2(88.5) = 372°\text{C}$. From Eq. (9.7), the maximum temperature at the tool rake face is

$$T_{\max} = T_o + T_s + T_m = 22 + 275 + 372 = 669°\text{C}.$$

Temperature on the newly machined surface T_w is

$$22 + T_w = 22 + \frac{\Gamma P_S}{\rho cvtb} = 22 + \frac{\Gamma}{1-\Gamma} T_S = 22 + \frac{0.1}{0.9}(275) = 52.5°\text{C}.$$

It should be noted that these calculations assume that the thermal properties of the material are constant and independent of temperature. In reality, with many engineering materials, the specific heat capacity and thermal conductivity do vary with changes in temperature. If more accurate predictions of cutting temperatures are required, the relationships between the thermal properties of the material and temperature must be known and taken into account.

Suppose the tool forces and the cutting ratio do not vary with cutting speed, for the conditions used in the case study the relationships between temperature and cutting speed as shown in Fig. 9.13 is obtained. Here, it is seen that the main shear-zone temperature increases slightly with cutting speed and then tends to constant out, whereas the maximum tool face temperature increases rapidly with cutting speed. It can be generally concluded that the cutting speed is the main operating variable that affects temperature. This is why the empirical relationship referred to as the Taylor's tool life equation depends mostly on speed.

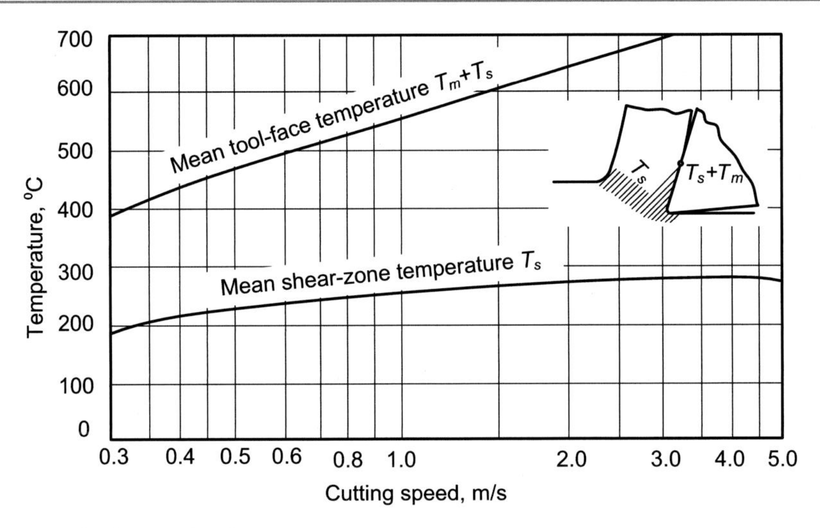

Fig. 9.13 Cutting speed and temperature

Homework

1. In a cutting process, the temperatures of the exposed chip surface and the newly machined workpiece surface are relatively easy to measure, while to obtain the temperature at the inside surface of the chip facing the tool rake is not trivial. With the use of the infrared technique in a shaping operation, the exposed chip surface showed 275 °C and the machined workpiece surface showed 87.5 °C. You are asked to estimate the maximum temperature at the inside surface of the chip in contact with the tool rake. The cutting conditions were known to be: tangential cutting force = 2500 N, cutting surface velocity = 1 m/s, chip thickness = 1.2 mm, undeformed chip thickness = 1 mm, width of chip = 2 mm, initial work temperature = 25 °C, length of contact between chip and tool face = 1.8 mm, workpiece density = 5000 kg/m^3, thermal conductivity = 62 W/m-K, and workpiece heat capacity = 600 J/kg-°C.
2. The specific cutting energy of a material is mostly a property-dependent quantity, and it can be identified in cutting based on the temperature distribution. For example, in the turning of a certain material, the temperature distribution is found to be 342 °C on the exposed chip surface, 602 °C at the chip–tool interface, and 74 °C on the machined surface. While the initial temperature of the workpiece is 22 °C, could you estimate its specific cutting energy (in the unit of N/mm^2)? Given that: thermal conductivity = 82 W/m-K, density = 6900 kg/m^3, specific heat capacity = 480 J/(kg)(°C), cutting velocity = 4.5 m/s, feed = 0.2 mm/rev, major cutting edge angle = 90°, radial depth of cut = 1 mm, chip thickness = 0.25 mm, contact length between chip and tool rake = 8 mm.
3. A magnesium alloy (density of 5800 kg/m^3, specific heat capacity of 105 J/kg(°C), and thermal conductivity of 75 W/m-K) is turned by a carbide tool of 10° rake angle and 90° major cutting edge angle. At a surface cutting velocity

of 8 m/s, feed of 0.15 mm, depth of cut of 0.15 mm, and a chip thickness of 0.45 mm, the temperature at the exposed chip surface is measured to be 320 °C and the maximum temperature at the chip–tool interface is 650 °C. If the cutting velocity is increased to 25 m/s, what will be the maximum temperature at the chip–tool interface? Note that the Weiner's solution for heat partition (as show in Fig. 9.9) can be approximated by $\Gamma = 0.47 - 0.385\log(R\tan\varphi)$. You can also assume that the secondary shear zone has zero width, that is, $w_o = 0$.

Machine Tool Chatter

10

10.1 Machining Dynamics

One of the important requirements that must be considered in designing machine tool structure and planning cutting process is vibration stability. In metal cutting, there is a possibility for the cutter to move in and out of the workpiece at an intensive magnitude that gives rise to undulations on the machined surface and excessive variations of the cutting force, which shorten the life of the cutter and the machine. Depending on the sources of energy, tool vibration can be generally categorized into two classes: forced vibrations and self-excited vibrations.

Forced vibrations during cutting are associated with disturbing periodic forces resulting from the unbalance of rotating parts, from errors of accuracy in some driving components, or simply from the intermittent engagement of workpiece with multi-flute cutters. The periodic forces supply the energy for forced vibrations, which result in waviness of the machined surface. If some of the harmonic components of the cutting force are in resonance with one of the significant modes of the machine structure, the resulting vibration can be rather intense. The machine design criterion is, thus, to increase the machine stiffness and to limit forced vibrations so as to achieve the required part quality.

Self-excited vibrations occur under certain conditions generally associated with the increase of the material removal rate, and these are energized by the cutting process itself. This class of vibration is often referred to as chatter, and it causes unacceptable machining finish and shortens the machine and tool life. Characteristic surface patterns caused by chatter are shown for turning and face milling in Fig. 10.1. It is required that the range of operations and of cutting conditions over which chatter occurs be avoided, and the cutting process must be stable at all times.

In Fig. 10.2, a typical record of force and vibration in milling is shown. These have been measured horizontally on the headstock of a horizontal machine when milling a steel workpiece of 8 in. width with a carbide cutter of 10 in. diameter with 12 teeth. The feed per tooth was 0.004 in., the depth of cut was 0.3 in., and the

S. Y. Liang, A. J. Shih, *Analysis of Machining and Machine Tools*,
DOI 10.1007/978-1-4899-7645-1_10

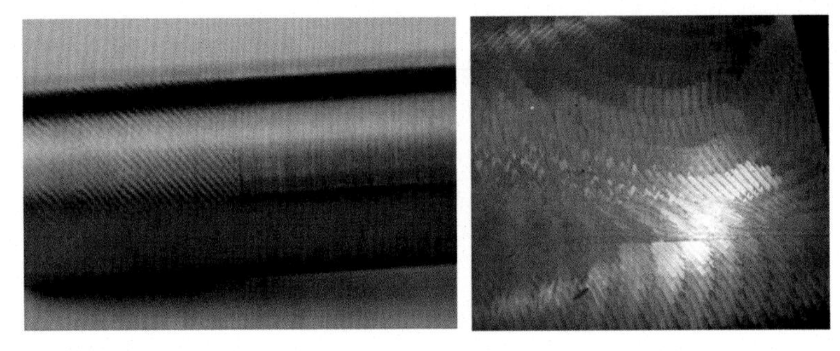

Fig. 10.1 Chatter marks in turning *(left)* and face milling *(right)*. (Source: Planlauf and Sherline)

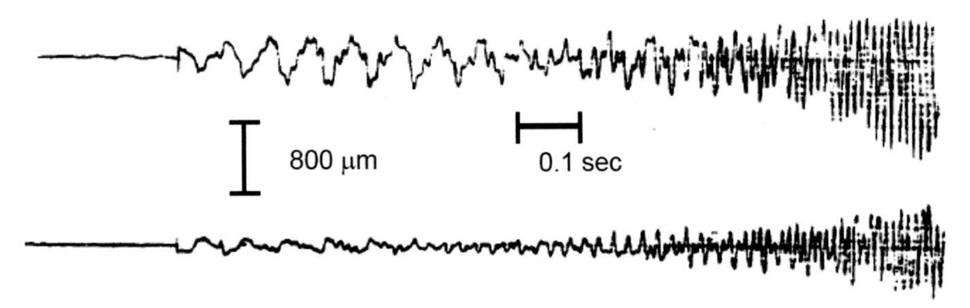

Fig. 10.2 Force *(top)* and displacement *(bottom)* in forced and self-excited vibration in face milling on a horizontal machine

spindle speed was 56 rev/min, which gave a tooth-passing frequency of 11.2 Hz. At the beginning of the cut, only forced vibrations at the tooth-passing frequency were measured. Before long, self-excited vibration developed with a frequency of 45 Hz; the amplitude of this vibration increased quickly, and the cut had to be interrupted. The amplitude of the self-excited vibration is seen to be higher than that of the forced vibration, and the frequency of the self-excited vibration was higher as it corresponded to the natural frequency of the machine structure. Also, the energy of the self-excited vibration was higher as it depended on the square of frequency. In the given case, the frequency of chatter was comparatively low. Usually, frequencies in the range 100–300 Hz are encountered.

The theory of chatter and stability analysis is the topic of discussion in this chapter. For simplicity, the discussion herein will be limited to the case of single-point turning process only. Two forms of chatter will be examined, all of which may be analyzed by simple linear system theory, although other nonlinear forms also exist, especially in view of the noncontinuous tool and work contact condition due to large vibration amplitude. The first form of chatter is called the Arnold-type chatter, named after the person who first described it systematically, which is due to the variation of force with cutting speed. The second form of chatter is usually known as regenerative chatter but it is sometimes associated with the names of Tobias and Tlusty, who described it independently. Regenerative chatter takes place as the cut-

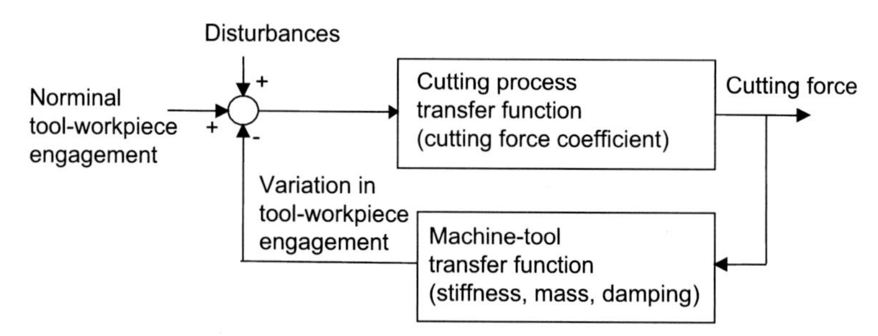

Fig. 10.3 Closed-loop machine tool dynamics

ting force varies as a result of unevenness of the surface being cut due to variations in the cutting force at the last time when the tool passed over that particular part of the surface.

All the forms of chatter can be understood as a feedback loop in the machine tool structure between the cutting process and the machine frame and drive system. Figure 10.3 shows the block diagram of such a loop. The transfer function of the machine tool, in terms of the stiffness and damping characteristics, is seen to play a critical role in the stability of the overall feedback system. Although machine tools vary so much not only between different types but even between successive tools of the same production line, one useful order of magnitude number is that the static stiffness of most machine tools measured between the cutting tool and the workpiece tends to be around 10^5 lbf/in (17.5 kN/mm). A number of 10^6 lbf/in is exceptionally good; one of 10^4 lbf/in. is rather bad, but may still be tolerable in certain low cost production with small machine tools.

10.2 Arnold-Type Chatter

This type of tool chatter takes place in a direction tangential to the workpiece as shown in Fig. 10.4. To examine the effect of this chatter type in isolation, it is considered that the machining system is rather compliant only in the cutting velocity direction. The workpiece and its drive system have relatively high rigidity and are rotating at a constant angular velocity. The tip of the tool can move in the plane of the paper, but is completely rigid in all other directions. In this case, the tool and its support mechanism may be regarded as a linear spring-mass damper system. Taking the effective mass of the system as m, the effective damping coefficient as b_x, and the effective static stiffness as k_x, the equation of motion may be written in the usual manner for a second-order dynamical system. Suppose the cutting force at any instant is given by $\overline{p} + p$, where $\overline{p}$ is the mean steady-state tangential cutting force and p is the instantaneous variation from it; and the instantaneous cutter displacement is given by $\overline{x} + x$, where $\overline{x}$ is the mean steady-state displacement and x is any deviation component, the static and dynamic conditions are

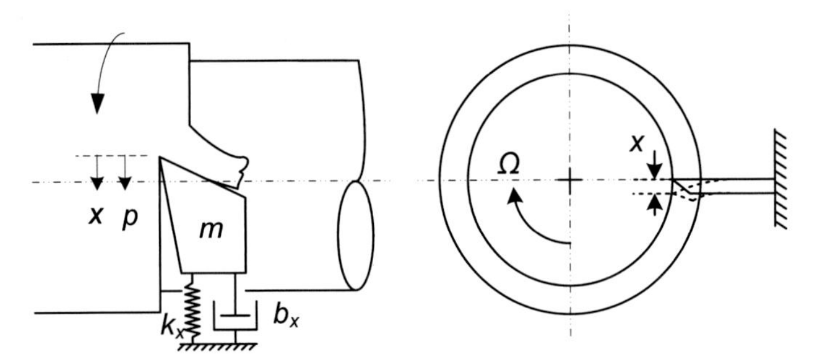

Fig. 10.4 Arnold-type chatter

$$\bar{x} = \frac{\bar{p}}{k_x} \tag{10.1}$$

$$m\ddot{x} + b_x\dot{x} + k_x x = p. \tag{10.2}$$

In terms of the undamped natural frequency ω_x and the damping ratio ξ_x given as follows

$$\omega_x = \sqrt{\frac{k_x}{m}} \text{ and } \xi_x = \frac{b_x}{2\sqrt{mk_x}}, \tag{10.3}$$

Eq. (10.2) can be rearranged into

$$\ddot{x} + 2\xi_x\omega_x\dot{x} + \omega_x^2 x = \frac{\omega_x^2}{k_x} p. \tag{10.4}$$

The tangential cutting force has been reported to be affected by the relative cutting velocity based on experimental data typically as seen in Fig. 10.5. As the tool moves downward with velocity variation of $\dot{x}$, the relative cutting velocity drops by an equal magnitude of $\dot{x}$. This implies that the cutting force increases with $\dot{x}$ according to Fig. 10.5. If the relationship between the force and the cutting speed in Fig. 10.5 is governed by a constant gradient λ, it follows that

$$p = -\lambda(\text{Cutting velocity}) = \lambda\dot{x}, \tag{10.5}$$

where λ assumes a positive value. Eliminating p from the above two equations gives

$$\ddot{x} + \left(2\xi_x\omega_x - \frac{\omega_x^2}{k_x}\lambda\right)\dot{x} + \omega_x^2 x = 0. \tag{10.6}$$

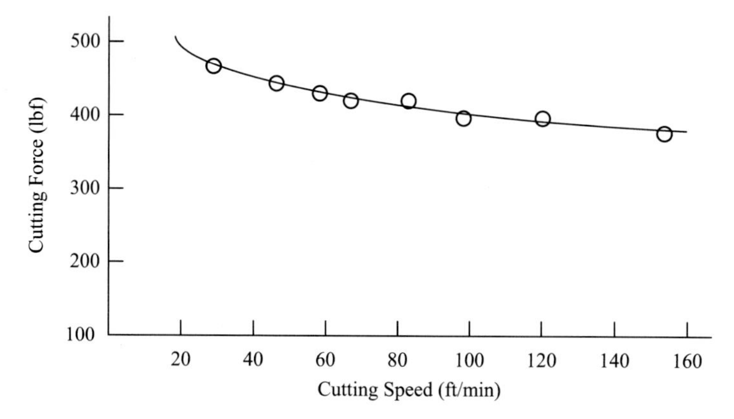

Fig. 10.5 Variation of cutting force with speed

Note that the damping ratio, ξ_a, of the system is now $\xi_x - \frac{\omega_x \lambda}{2k_x}$ associated with the coefficient of the first-order term.

Oscillatory motion will appear in response to any initial disturbance that triggers a tool-work relative motion if the magnitude of the damping ratio is less than 1:

$$\left| \xi_x - \frac{\omega_x \lambda}{2k_x} \right| < 1 \Rightarrow -1 < \xi_x - \frac{\omega_x \lambda}{2k_x} < 1, \tag{10.7}$$

which is a condition often satisfied in reality. Equation (10.6) also suggests that unstable (chatter) motion will take place, while any initial disturbance will turn into increasing and eventually unbounded tool displacement, if and only if

$$\xi_x - \frac{\omega_x \lambda}{2k_x} \leq 0 \Rightarrow \xi_x \leq \frac{\omega_x \lambda}{2k_x} \tag{10.8}$$

or, with the substitution of Eq. (10.3), simply

$$b_x \leq \lambda. \tag{10.9}$$

In practice, the unstable tool motion causes the tool to disengage from the workpiece at large displacement, and the linear equation of motion begins to fail. The nonlinearity determines the upper bound of the amplitude of the tool chatter motion. Although λ is an interface-dependent quantity, Eq. (10.9) implies that increasing the damping of the machine increases its stability.

Example

Explain how you can estimate, off-line, the frequency of the Arnold-type chatter vibration in a turning operation with an infinitely rigid drive–spindle–workpiece assembly.

Solution:

The observed (damped) frequency of the Arnold-type chatter is

$$\omega_x\sqrt{1-\xi_a^{\ 2}} = \omega_x\sqrt{1-\left(\xi_x - \frac{\omega_x\lambda}{2k_x}\right)^2}.$$

Therefore, an off-line model testing can be used in conjunction with the knowledge on λ (which can be found experimentally by measuring forces in various cutting speeds).

Example

The following figure shows the cutting force in a turning operation at various speeds with an aluminum alloy workpiece. Consider a machine tool that has a natural frequency of 650 Hz and a damping ratio of 0.35 along the workpiece tangential direction according to the data from an impulse response test. What is the minimum stiffness that this machine ought to have to avoid chatter of Arnold type when cutting the aluminum alloy workpiece?

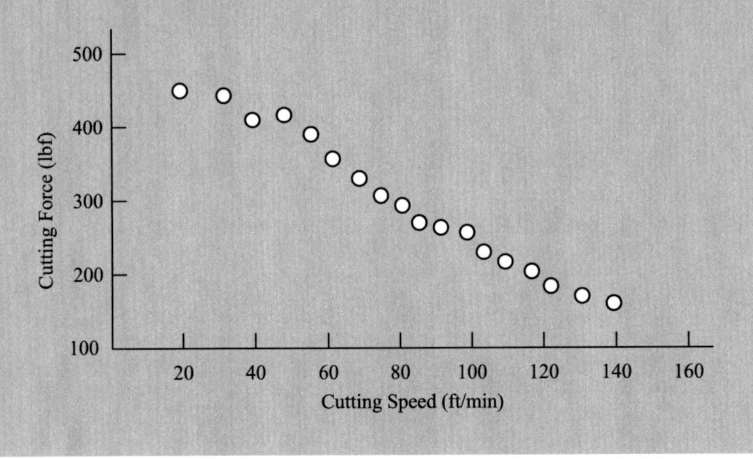

Solution:

$$\lambda \approx 2.27\,\text{lb-min/ft.}=11.36\,\text{lb-s/in.}$$

To avoid the Arnold-type chatter: $b_x = \dfrac{2\xi_x k_x}{\omega_x} > \lambda \Rightarrow \dfrac{2(0.35)k_x}{650} > 11.36$

$$\Rightarrow k_x > 10{,}550\ \text{lb/in.}$$

The above one-degree-of-freedom analysis assumes that the workpiece–spindle assembly is perfectly rigid. This assumption can be relaxed to take note of the variations in angular velocity of the work due to the flexibility of the drive system. If at any moment the position of the workpiece is given by $\bar{\theta}+\theta$, where $\bar{\theta}$ is the nominal position and θ the variational component, Eq. (10.5) is modified to

$$p=\lambda(\dot{x}-R\dot{\theta}), \tag{10.10}$$

where R is the mean radius of the workpiece. If the drive system combined with the workpiece has an undamped natural frequency ω_θ, damping ratio ξ_θ, and stiffness k_θ, then

$$\ddot{\theta}+2\xi_\theta\omega_\theta\dot{\theta}+\omega_\theta^2\theta=\frac{\omega_\theta^2}{k_\theta}(k_e\dot{\theta}-Rp). \tag{10.11}$$

The most common form of machine-tool driving motor is an induction motor, which provides a torque roughly proportional to the spindle speed. The proportional constant is given as k_e in the above equation.

Transforming Eqs. (10.4), (10.10), and (10.11) into the Laplace domain gives the following expressions

$$\left(s^2+2\xi_x\omega_x s+\omega_x^2\right)X(S)=\frac{\omega_x^2}{k_x}(P(s)) \tag{10.12}$$

$$P(s)=\lambda\left[sX(s)-Rs\Theta(s)\right] \tag{10.13}$$

$$\left(s^2+2\xi_\theta\omega_\theta s+\omega_\theta^2\right)\Theta(s)=\frac{\omega_\theta^2}{k_\theta}\left(k_e s\Theta(s)-RP(s)\right). \tag{10.14}$$

These equations collectively govern the dynamics of the entire workpiece–tool system. The two degrees of freedom are coupled through the cutting force given in Eq. (10.13). Eliminating $P(s)$ and $\Theta(s)$ from the above gives:

$$(a_4s^4+a_3s^3+a_2s^2+a_1s+a_0)X(s)=0 \tag{10.15}$$

with $a_4=\dfrac{1}{\omega_\theta^2\omega_x^2}$, $a_3=\dfrac{2\xi_\theta}{\omega_\theta\omega_x^2}+\dfrac{2\xi_x}{\omega_\theta^2\omega_x}-\dfrac{k_e}{k_\theta\omega_x^2}-\dfrac{R^2\lambda}{k_\theta\omega_x^2}-\dfrac{\lambda}{\omega_\theta^2k_x}$,

$$a_2=\frac{1}{\omega_\theta^2}+\frac{1}{\omega_x^2}+\left(\frac{2\xi_\theta}{\omega_\theta}-\frac{k_e}{k_\theta}-R\frac{\lambda}{k_\theta}\right)\left(\frac{2\xi_x}{\omega_x}-\frac{\lambda}{k_\theta}\right)-\frac{\lambda R^2}{k_\theta k_x},$$

$$a_1=\left(\frac{2\xi_\theta}{\omega_\theta}-\frac{k_e}{k_\theta}-\frac{R^2\lambda}{k_\theta}\right)\frac{2\xi_x}{\omega_x}-\frac{\lambda}{k_x}\ \text{and}\ a_0=1.$$

The above fourth-order differential equation represents the motion of x, which may be stable or unstable depending on the roots of the equation. Routh criterion can usually be used to determine the stability of the system given the parameters a_1, a_2, a_3, and a_4. The added workpiece flexibility contributes significantly to the analysis complexity of the system dynamic behavior.

10.3 Regenerative Chatter

This type of tool chatter takes place as a result of non-uniform (uncut) chip thickness due to variations in the cutting force on the last revolution, when the tool passed over that particular part on the surface. In Fig. 10.6, the cutter is being fed to the left at a steady speed. It is considered as having a static stiffness of k_y and damping coefficient of b_y in the y direction and rigid in all other directions. At any moment, because of the cutting action, a force $\bar{f}+f$, consisting of a mean component $\bar{f}$ and a variational component f, is applied on the tool in the y direction. Meanwhile, the displacement of the tool can be described in terms of a mean component $\bar{y}$ and a variational component y.

There is a well-accepted assumption that the cutting force is proportional to the instantaneous thickness of the chip being cut, and the following analysis will be based on this assumption. The instantaneous chip thickness is determined by the present position of the tool as well as by the position of the tool one revolution previously. Considering only the variational components, one can relate the force on the tool to the displacement by the equation:

$$f = k_t\left(y\left(t-\tfrac{2\pi}{\Omega}\right)-y(t)\right), \tag{10.16}$$

where Ω is the constant angular velocity of the workpiece and k_t is usually referred to as the cutting force coefficient and it has the unit of stiffness.

In the Laplace domain, the above equation has the form of

$$F(s) = k_t\left(e^{-2\pi s/\Omega}-1\right)Y(s). \tag{10.17}$$

Similar to Eq. (10.4), the equation of motion of the tool in the y direction is

$$\ddot{y}+2\xi_y\omega_y\dot{y}+\omega_y^2 y = \frac{\omega_y^2}{k_y}f, \tag{10.18}$$

where $\omega_y = \sqrt{\dfrac{k_y}{m}}$ and $\xi_y = \dfrac{b_y}{2\sqrt{mk_y}}$. Taking the Laplace transform of the above equation of motion and substituting $F(s)$ with that in Eq. (10.17) gives

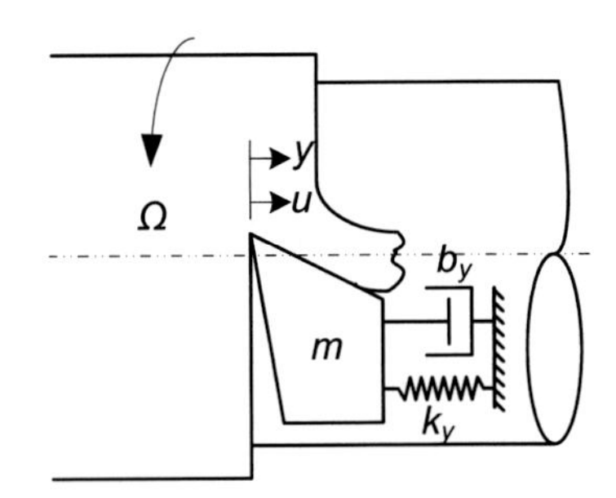

Fig. 10.6 Regenerative chatter

$$\left\{e^{\frac{2\pi}{\Omega}s}\left[\frac{s^2}{\omega_y^2}+\frac{2\xi_y}{\omega_y}s+\left(1+\frac{k_t}{k_y}\right)\right]-\frac{k_t}{k_y}\right\}Y(s)\equiv G(s)Y(s)=0. \tag{10.19}$$

The roots of $G(s)$ in the above equation have a very simple physical meaning in the context of chatter stability. If the roots of $G(s)$ are purely imaginary, in the form of $\pm j\omega$, the free response of $y(t)$ to a nonzero initial displacement will neither converge nor diverge. In the case of $\omega \neq 0$, the response will be oscillatory with a frequency of ω and a constant amplitude equivalent to the initial displacement. This is a condition described as limitedly stable, and it is the boundary that marks chatter stability. In metal cutting, initial displacements of the tool can be attributed to mechanical disturbance resulting from nonuniform material properties, variation of machine performance, or the change of machine–workpiece interface condition. In addition to the efforts of eliminating disturbance sources, the important criterion for machine tool design is to avoid unstable vibration modes, assuming the disturbances will always be present.

Defining $H(s)$ as

$$H(s)=\frac{s^2}{\omega_y^2}+\frac{2\xi_y}{\omega_y}s+\left(1+\frac{k_t}{k_y}\right) \tag{10.20}$$

Then, the chatter stability boundary as given by the following condition

$$G(s)\,|_{s=j\omega}=0 \tag{10.21}$$

can be written as

$$\left[e^{\frac{2\pi}{\Omega}s}H(s)\right]_{s=j\omega}=\frac{k_t}{k_y}. \tag{10.22}$$

Equation (10.22) can be satisfied if all the following conditions Eqs. (10.23)–(10.25) are met simultaneously:

$$\left|H(s)|_{s=j\omega}\right|=\frac{k_t}{k_y} \tag{10.23}$$

$$\angle\left(H(s)|_{s=j\omega}\right)=\frac{\pi}{2} \tag{10.24}$$

$$\angle e^{j2\pi\omega/\Omega}=\frac{3\pi}{2}+2n\pi,\ \text{with n}=0,1,2,\ldots \tag{10.25}$$

Also, note that these are sufficient but not necessary conditions for chatter. From the definition of Eq. (10.20),

$$H(s)|_{s=j\omega}=\left(1+\frac{k_t}{k_y}\right)-\frac{\omega^2}{\omega_y^2}+j\frac{2\xi_y}{\omega_y}\omega\equiv \text{Re}+j\,\text{Im}. \tag{10.26}$$

If Eq. (10.24) is true, Re=0 and it follows that

$$\frac{\omega^2}{\omega_y^2} = 1 + \frac{k_t}{k_y} \Rightarrow \omega = \omega_y \sqrt{1 + \frac{k_t}{k_y}}. \tag{10.27}$$

If Eq. (10.23) is true in addition,

$$\mathrm{Im} = \left| H(s)\big|_{s=j\omega} \right| = \frac{k_t}{k_y}, \tag{10.28}$$

which suggests that

$$\frac{2\xi_y \omega}{\omega_y} = \frac{k_t}{k_y} \Rightarrow \omega = \left(\frac{k_t}{2\xi_y k_y} \right) \omega_y. \tag{10.29}$$

Combining the above with Eq. (10.27) gives

$$\frac{k_t}{2\xi_y k_y} = \sqrt{1 + \frac{k_t}{k_y}}. \tag{10.30}$$

With most of the machine tools, k_y is an order of 10^5 lb/in., k_t is found to be in the neighborhood of 2×10^3 to 5×10^4 lb/in., and ξ_y is to be about 0.01–0.25.

For condition Eq. (10.25) to be met,

$$\frac{\Omega}{\omega} = \frac{4}{3 + 4m}, \text{ with } m = 0, 1, 2, \ldots \tag{10.31}$$

This implies, as is found in practice, that the sinusoidal irregularity of the surface of the workpiece moves circumferentially relative to the workpiece.

The solutions for the stability condition in Eq. (10.22) are graphically shown in Fig. 10.7. The lobes in the figure represent the stability boundaries containing shaded regions of chatter that can take place over a range of cutting speed. Note that Eqs. (10.23)–(10.25) do not provide necessary and sufficient conditions for chatter and they determine only the lowest points on the lobes.

To avoid chatter from Fig. 10.7, it is seen that a sufficient condition is

$$\frac{k_t}{2\xi_y k_y} < 1. \tag{10.32}$$

The basic definition of damping ratio ξ_y and natural frequency ω_y gives the following

$$b_y \omega_y = 2\xi_y k_y. \tag{10.33}$$

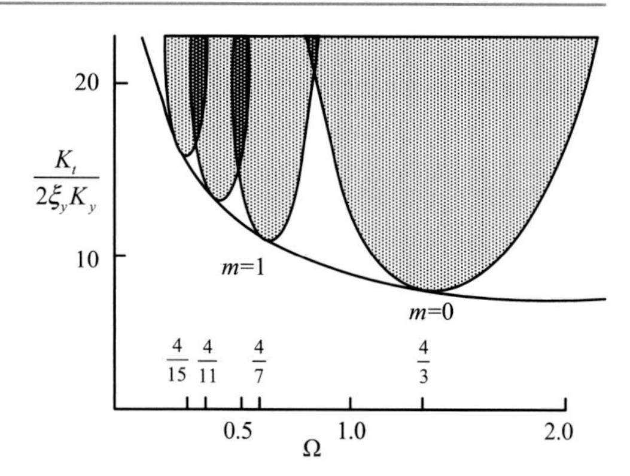

Fig. 10.7 Stability lobes with instability in shaded portions

Combining Eqs. (10.32) and (10.33) leads to a simple and very useful guideline for the design of machine tools against chatter:

$$b_y \omega_y > k_t. \tag{10.34}$$

In other words, the machine resonant dynamic stiffness should always be made large enough to overcome the cutting pressure constant of the material.

All other conditions unchanged, increasing the damping of the machine tool will reduce the tendency of a machine to chatter, whether the chatter be Arnold or regenerative type. Increasing the stiffness reduces the chance of regenerative-type chatter but often runs the risk of reducing the damping due to possible elimination of joints. Thought may need to be given as to how to ensure sufficient damping, although increasing the stiffness of the machine tool has the further advantage that more accurate product can be expected under all cutting conditions.

The theory is inadequate for the reason that only one mode of vibration has been considered; where other modes exist with very similar frequencies, coupling can occur, and the tool may have motions along three orthogonal axes and also torsionally about its axes. Nevertheless, the elementary theory covered in this chapter gives a clear guide to the machine designer; in particular, it draws attention to the importance of raising the stiffness of the machine.

Example

With an experimental modal analysis, the stiffness, natural frequency, and damping ratio of an engine lathe have been identified to be 10^5 lb/in., 150 Hz, and 0.2, respectively, along the spindle axis direction. If a cylindrical aluminum alloy is to be turned at 4×10^{-3} ipr, what is the limiting radial depth of cut beyond which tool chatter may take place? Note that the following specific cutting energy curves can be useful if you recall that [cutting force = (specific cutting energy) × (chip cross sectional area)]. 1 J = 16.2 lbf-in.

Solution: From Fig. 2.4 at an undeformed chip thickness of 0.004 in (0.1 mm), specific cutting energy u $= 1.2$ GJ/m^3 $= 3.2\times10^5$ lbf/in.^{2}

Since $f = ub\Delta y$ and $f = k_t\Delta y$ (Eq. 10.16), $k_t = ub$.

The condition to prevent regenerative chatter is $b_y\omega_y > k_t$, which is then the same as $b < \frac{b_y\omega_y}{u}$.

Note that $b_y = \frac{2\xi_y k_y}{\omega_y} = \frac{2(0.2)(10^5)}{150} = 266.6\ \frac{\text{lb-s}}{\text{in.}}$.

Therefore, $b < \frac{b_y\omega_y}{u} = \frac{(266.6)(150)}{3.2\times10^5} = 0.125$ in.

Homework

1. With off-line impact testing and force-displacement testing, it has been found that the dominant dynamic mode of a certain lathe has a damping ratio of 1.227, natural frequency of 800 Hz, and stiffness of 20,000 lb/in. When carbide inserts are used on this lathe to machine medium carbon steel, it is observed that stable Arnold-type vibration takes place occasionally due to the impurity of the workpiece material. The vibration has a period of about 1.89×10^{-3} s. If the machine stiffness could be improved, what should be the minimum stiffness required to eliminate the vibration?
2. A 5° rake cutter is used to turn a medium carbon steel (apparent shear strength τ of 300 kpsi) bar. The friction coefficient between the tool–workpiece interface is 0.25. From its response to a hammer impact, the machine tool dynamics can be approximated by a second-order system with 7.6×10^{-4} s of time constant and 816.5 Hz of damped natural frequency. At a radial depth of cut of 0.1 in., what is the minimum acceptable stiffness of the machine tool to avoid regenerative chatter? Note that $F_P = \frac{\tau bt}{\sin\varphi}\left(\frac{\cos(\beta-\alpha)}{\cos(\varphi+\beta-\alpha)}\right)$.
3. Insert realistic numbers in Eq. (10.15) and run the Routh test.

11 Electrical Discharge Machining

The machining processes introduced in the previous chapters focus on the generation of relative motion between the workpiece and the wedge-shaped cutting tool. In these traditional machining processes, the fundamental mechanism responsible for the removal of materials is the plastic shearing action along the shear plane. However, these processes are effective only when the tool material is harder than the work material, which exhibits reasonable amount of plasticity prior to the occurrence of gross fracture. It is also a prerequisite to the success of these processes that workpiece materials have lower shear strength or wear resistance than that of the tool materials.

For machining of many hard and brittle materials, such as high-strength heat-resistant alloys, hardened steels, fiber-reinforced composites, ceramics, metal matrix ceramic composites, or soft elastomers and tissues, the traditional processes can no longer be used effectively. There are a number of processes that remove materials by melting, evaporation, chemical action, electrical energy, and/or hydraulic power, collectively referred to as the nontraditional (or unconventional) machining processes. These processes are not affected by the hardness and brittleness or the softness and large deformation of the workpiece materials. Some of these processes can be applied to fully heat-treated materials to produce a net-shape part, thus finding applications in high demand in aerospace, automotive, electronics, and die and mold industries. The others are widely used in healthcare during clinical procedures such as the insertion of needles for biopsy cutting of tissue samples, needle guidance to reach specific sites for delivery of therapy or placement of radioactive seeds, and electrosurgical cutting and coagulation of the tissue. In surgery and needle-based minimally invasive procedures, machining is the fundamental process and tissue is the workpiece.

Nontraditional machining processes include electrical discharge machining (EDM), electrochemical machining, chemical machining, electron beam machining, ultrasonic machining, water-jet machining, and many others.

S. Y. Liang, A. J. Shih, *Analysis of Machining and Machine Tools*,
DOI 10.1007/978-1-4899-7645-1_11

11.1 EDM—An Overview

EDM process utilizes sparks generated between the electrode and workpiece to melt and remove the work material and create the desired geometry and surface integrity. It is one of the earliest nontraditional machining processes that has been developed and applied extensively in the industry for machining of micro features and difficult-to-machine materials.

EDM was first developed by two Russian engineers, B. R. Lazarenko and N. I. Lazarenko, in 1943. They applied the commonly observed phenomenon of erosion on electrical switch to machining. By controlling the gap distance, using suitable dielectric medium, and delivering a series of voltage pulses between the electrode and workpiece, a series of electrical sparks can be generated by pulsing the electrical power input. The spark creates high temperature instantaneously and removes the work material and subsequently creates wear on the electrode.

EDM has several advantages over the traditional machining processes, which rely on a much harder tool to remove the work material by generating chips. First, the hardness of the electrode and work materials is not a factor in machinability. Since melting is the primary mechanism for material removal in EDM, the melting temperature and heat capacity of work and electrode materials are more important in determining the material removal rate (MRR). EDM can be applied to machine the hard, difficult-to-cut materials, such as hardened steels, tungsten carbide, polycrystalline diamond, and polycrystalline cubic boron nitride. Second, the cutting force created by EDM is small. This makes EDM an ideal process for machining miniature features. For example, the spray holes in diesel fuel injector with diameters ranging from 80–200 μm are produced by EDM using a thin, long wire electrode. Third, EDM can generate complicated and precise geometry. The electrode can be machined to the shape and its form is transferred to the workpiece, or a thin wire running in the axial direction can be tilted at varying angles when the electrode cuts the workpiece and creates micro features with intricate and accurate geometry. EDM process has been widely utilized in the fuel injector for gasoline and diesel engines, turbine blade for aerospace and power generation, precision cutting and molding tools, etc.

EDM also has several limitations. The work material prefers to be electrically conductive. The electrode wears during EDM and affects the part accuracy. Spark erosion is generally a slow process. The MRR is typically low. High MRR EDM is possible in some configurations, such as the EDM hole drilling using a rotary tubular electrode with high pressure dielectric medium delivered from inside the tube for flushing and the use of oxygen as the dielectric medium to enhance the energy generation and increase the MRR. Research has also been conducted in EDM of non-electrically conductive materials. This is possible but the MRR is very low. The electrode wears in EDM because the high temperature spark erodes both the workpiece and electrode. To improve part accuracy, compensation of electrode wear or use of a new tool for finishing is commonly required in EDM.

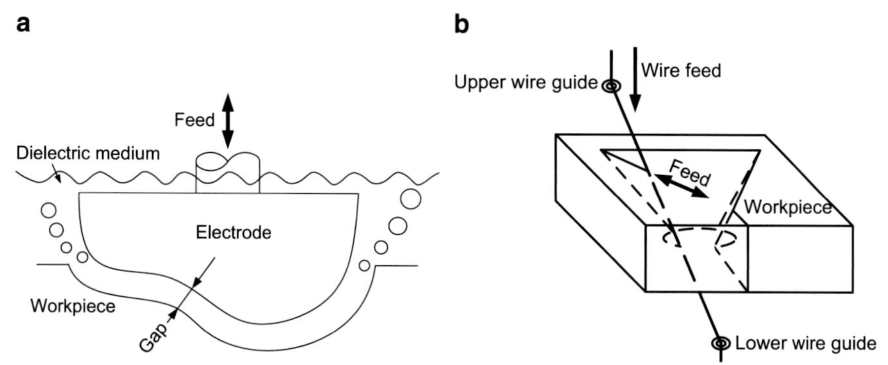

Fig. 11.1 A schematic illustration of the EDM process. **a** Sinking EDM and **b** wire EDM

11.2 Sinking and Wire EDM

EDM can be generally categorized into sinking and wire EDM. Sinking, also commonly known as die-sinking EDM, as shown in Fig. 11.1a, utilizes the shaped electrode to generate the opposite shape on the workpiece. The electrode is machined or formed to create the desired shape which is transferred to the workpiece. By controlling the relative motion of the electrode and workpiece, sinking EDM has several different configurations. For example, by orbiting (rotating and translating) using a thread as the electrode, the inner thread geometry can be generated in hardened tool steel and tungsten carbide. Another example uses a rotating thin wire feeding into the workpiece to drill high aspect ratio microholes can be generated. This is a particularly important process for the machining the microholes for diesel engine fuel injector. The spray hole geometry controls the spray pattern and is critical to the diesel engine fuel efficiency and emission performance. By rotating a tubular electrode and controlling its position and orientation, an electrode with simple geometry could be used like an end milling tool for EDM milling of sculpture surface.

Wire EDM uses a thin wire running in the axial direction, as shown in Fig. 11.1b. The orientation of the electrode is controlled by the upper and lower wire guides, usually made of diamond, to change the wire orientation to machine parts with complex shape and tight tolerances on workpiece. Voltage is delivered to the wire electrode by a metal contact, typically made of wear-resistance tungsten carbide. Very thin wire, as small as 30 μm in diameter, has been developed for wire EDM machining of micro features.

11.3 Discharge Material Removal Mechanism and Dielectric Medium

Discharging sparks in the gap between the electrode and workpiece create heat and melt the work material. In EDM, as shown in Fig. 11.2, the workpiece is commonly set as the anode and the electrode/tool as cathode. This setup is often called the

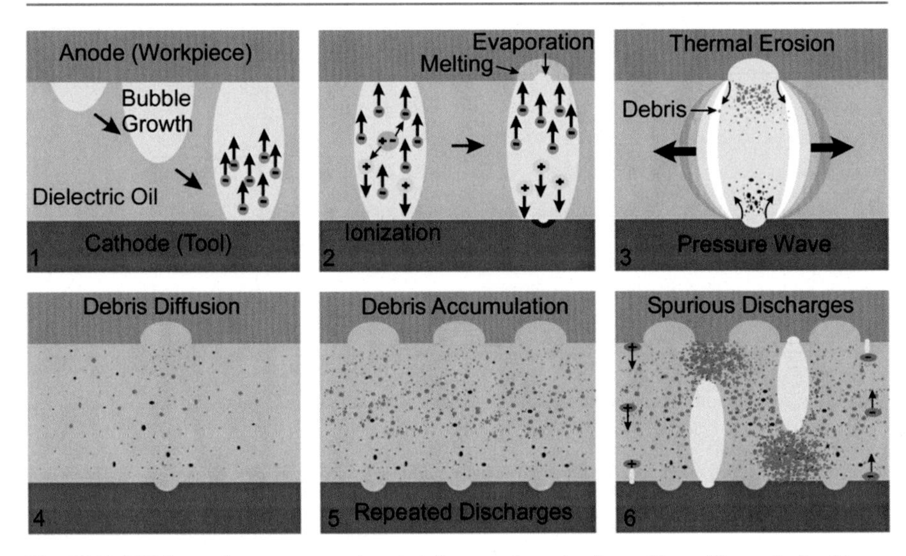

Fig. 11.2 EDM gap phenomenon and material removal mechanism with positive polarity; Stages 1–4: normal discharges; Stage 5: repeated discharges leading to debris accumulation; and Stage 6: excess debris causing spurious discharges through clusters of debris *(middle)* as well as by individual debris particles

positive polarity. This polarity setup generally has a low wear rate of the electrode (cathode) and high MRR of the workpiece.

When the gap is smaller than a critical value and a sufficient voltage is applied between the electrode and workpiece, sparks occur in the gap. A normal discharge is accompanied by the electron migration, dielectric medium ionization, electron avalanching, melting and evaporation, debris generation, and expansion of the bubble column with a pressure wave front. There are illustrated in Stages 1–4 in Fig. 11.2. The debris is diffused in the dielectric medium and accumulated after repeated discharge (Stage 5). Excess debris causes the spurious discharges through clusters of debris and individual debris particles (Stage 6).

In the gap between the electrode and workpiece lies the dielectric medium. The electrical, mechanical, and thermal properties of the dielectric fluid influence the processes of discharge initiation, plasma expansion, material removal, debris flushing, and discharge channel reconditioning in EDM. Table 11.1 summarizes key properties of the gas and liquid as EDM dielectric fluids. The dielectric strength determines the gap distance between the electrode and workpiece. The dielectric strength is defined as $\xi = u_s / h_g$. Higher dielectric strength ξ requires a higher electric voltage u_s to breakdown the dielectric fluid, and decreases the gap distance h_g. The dielectric constant determines the stray capacitance induced by the overlapping area between the electrode and workpiece. Large inertia and high viscosity of the dielectric fluid increase the bubble expansion force and material removal per discharge, which commonly corresponds to a higher MRR and a rougher surface finish. Heat conductivity and heat capacity are important factors that affect the solidification of molten debris and the cooling of the electrode and the workpiece surfaces.

Table 11.1 Electrical, thermal, and mechanical properties of liquid and gas dielectric media at room temperature

	Liquid		Gas		
	Deionized water	Kerosene/ hydrocarbon oil	Air	N_2	O_2
Dielectric strength ξ (MV/m)	13	14–22	3.0	2.8	2.6
Dielectric constant	80	1.8	1.0	1.0	1.0
Dynamic viscosity (g/m-s)	0.92	1.6	0.019	0.017	0.020
Thermal conductivity (W/m-K)	0.61	0.15	0.026	0.025	0.026
Heat capacity (J/g-K)	4.2	2.2	1.0	1.0	0.92

11.4 EDM Pulses and Process Monitoring

EDM process has a series of discharge pulses between the electrode and workpiece. As shown in Fig. 11.3, EDM pulses can be classified as the spark, arc, and short pulses by examining the voltage and current signal across the gap between the electrode and workpiece. The spark, as illustrated in Fig. 11.3a, occurs after the voltage reaches a specific level V_h. At the point of discharging, the voltage drops and current rises quickly to I_h. After the discharge the voltage oscillates, which is called the ringing effect, and a reversed current is typically observed. Figure 11.3b shows the voltage and current for an arc pulse, which has no ignition delay because the deionization from the previous pulse is not complete and the remnant plasma channel has a residual conductivity that provides a path for electric current. The peak voltage has not reached the preset value V_h but is still above a threshold value V_l.

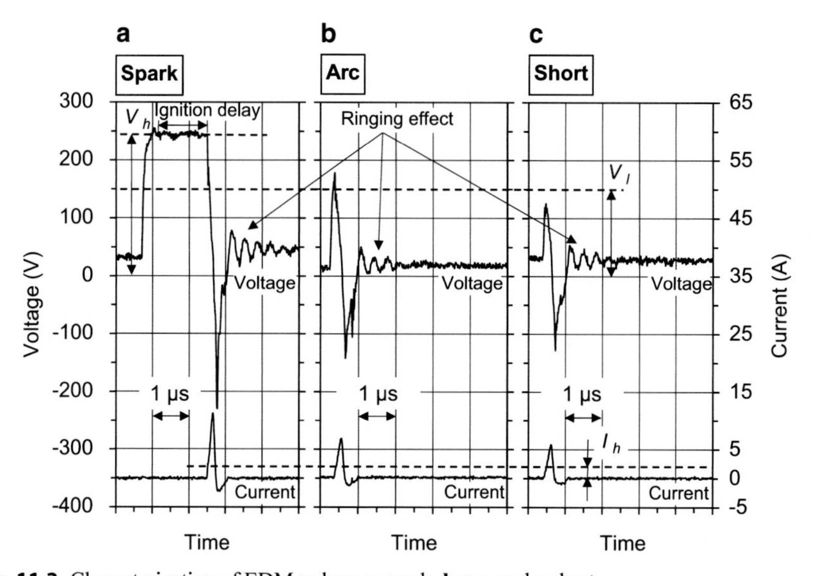

Fig. 11.3 Characterization of EDM pulses: **a** spark, **b** arc, and **c** short

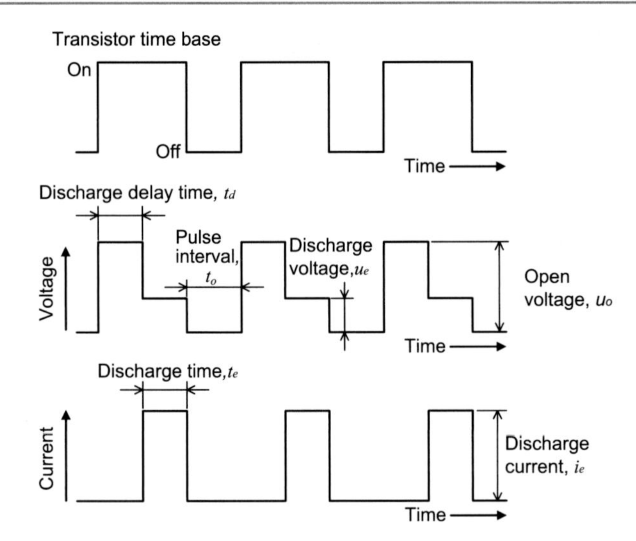

Fig. 11.4 Pulse chain and six key EDM process parameters

Short circuit occurs when the electrode contacts the workpiece. The voltage is lower than V_l at the peak of current, as shown in Fig. 11.3c.

The voltage and current signals in EDM can be utilized to monitor the EDM voltage and current waveform and determine the type of pulses and frequency of sparks. Six key EDM process parameters for most EDM machines, as illustrated in Fig. 11.4 with the ideal pulse waveform generated by an ideal transistor for on and off switching, are:

- Open voltage (u_o)*:* The voltage when the EDM circuit is in the open state and energy has been built up for discharging.
- Discharge voltage (u_e)*:* The voltage during discharge.
- Discharge current (i_e)*:* The current during discharge.
- Discharge delay time (t_d)*:* The time duration when the circuit is energized to open voltage and waiting for discharge.
- Discharge time (t_e)*:* The time duration for discharge.
- Pulse interval (t_o)*:* The waiting time to be energized to open voltage.

These parameters are determined by the EDM generator, dielectric medium, work and electrode materials, flushing conditions in the gap, and process parameter setup.

11.5 EDM Generators

EDM voltage and current are controlled by the generator to start and maintain the electric discharging process. Two main types of generator are the resistance-capacitance (RC) and transistor-based EDM pulse generators.

The RC circuit is basically a relaxation oscillator with a resistor and a capacitor, as illustrated in Fig. 11.5a. It is a simple, reliable, robust, low-cost power source for EDM. It can provide very small pulse energy and is used extensively in micro EDM and finishing EDM to achieve fine surface finish. The drawback of RC generator is the lack of precision control, particularly for timing and slow charging.

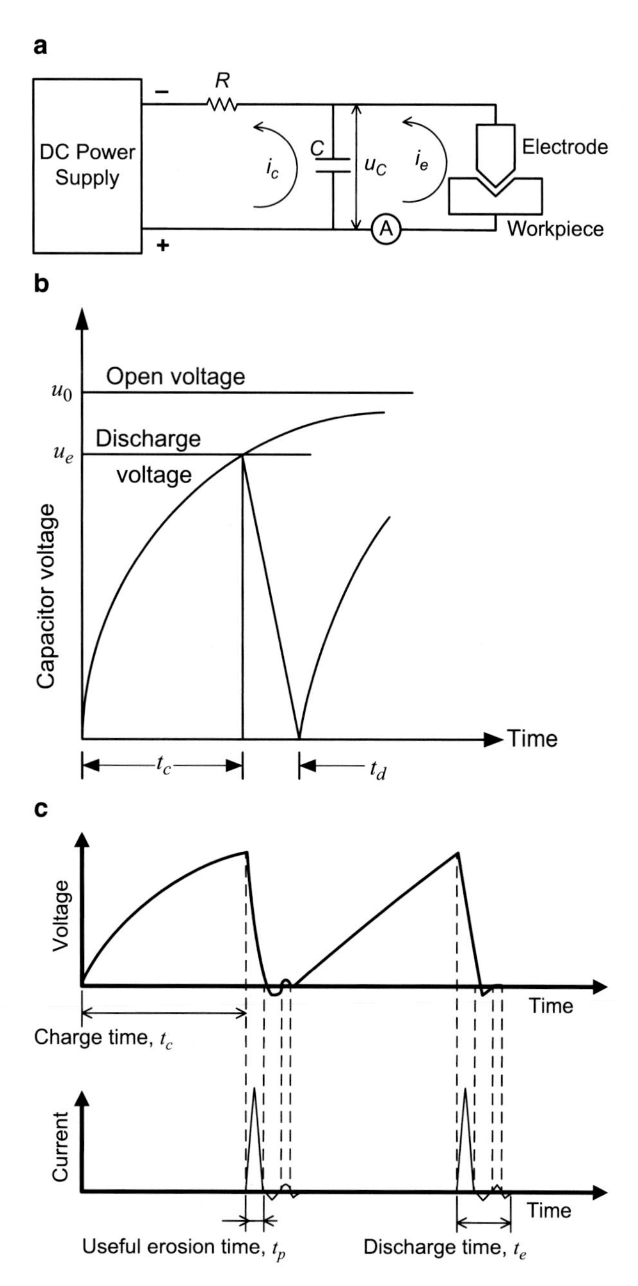

Fig. 11.5 RC EDM generator **a** overview of the circuit, **b** charging of the voltage between the electrode and workpiece, and **c** repeated charging and discharging

Figure 11.5b shows a typical voltage versus time curve for charging and discharging of an RC circuit. In the beginning, the capacitor is charged by a DC power source with open voltage u_o. The resistance R (in the unit of Ω) and capacitance C (in the unit of Farad) determine the charging rate and the voltage between the electrode and workpiece:

$$u_c = u_o(1 - e^{-t/RC}), \tag{11.1}$$

where t is the charging time. When u_c reaches the breakdown level, u_e, the discharge spark occurs. The energy E_d of each individual spark discharge is:

$$E_d = {}^1/_2\, Cu_e^2. \tag{11.2}$$

The cycle charging and discharging is repeated (Fig. 11.5c) until the cut is performed. The cycle frequency f_r is:

$$f_r = 1/(t_c + t_e), \tag{11.3}$$

where t_c is the charge time and t_e is the discharge time.

One strategy for the RC generator to improve the MRR is by reducing the time constant RC to give rapid charging and more frequent discharge. However, the frequency may reach a limit at which arcing occurs. Arcing creates surface thermal damages and slows the MRR. For maximum MRR, a study by Barash has provided the guideline that $u_e = 0.73u_o$ and $t_e = 0.1t_c$ are the preferred setup conditions. This guideline will be utilized in the following examples.

Example

In an EDM operation using RC generator, $u_o = 250\ V, R = 10\ \Omega,\ and\ C = 3\ \mu\text{F}$.
Under the maximum MRR condition, calculate:

(a) Discharge voltage, u_e
(b) Charging time, t_c
(c) Cycle frequency, f_r
(d) Energy of individual discharge, E_d
(e) If the dielectric strength ξ is 180 V/25 μm, estimate the gap distance

Solution

(a) The EDM is performed at maximum MRR condition using the RC generator with $u_e = 0.73u_o = 0.73 \times 250 = 182.5$ V
(b) From Eq. (11.1), at discharge $182.5 = 250(1 - e^{-t_c/30})$. The charging time $t_c = 39.3\ \mu$s (C in μF, and t_c in μs)

(c) At maximum MRR, discharge time $t_e = 0.1t_c = 3.93$ s. $t_c + t_e = 39.3 + 3.93 = 43.2$ μs. The cycle frequency, $f_r = 1/(t_c + t_e) = 23{,}100$ Hz = 23.1 kHz
(d) From Eq. (11.2), the energy of individual discharge, $E_d = {}^1/_2 C u_e^{\,2} = 0.5 \times 3 \times 10^{-6} \times 182.5^2 = 0.05$ J
(e) The gap distance, $h_g = u_e/\xi = 25.3\ \mu$m

Power electronics has revolutionized the commercial EDM generator as the mainstream technology. Modern power electronics using the metal-oxide-semiconductor field-effect transistor (MOSFET) technology can switch the electrical current and voltage on and off at high frequency. It is widely used in commercial EDM machines. The transistor EDM generator, as illustrated in Fig. 11.6, has the flexibility and performance advantages over the traditional RC generator in two ways. First, unlike the RC generators that produce isofrequency pulses (Fig. 11.7a), the transistor-based generator enables the isopulse (Fig. 11.7b) to control the duration of discharge to enhance the consistency in flushing the debris and the MRR. Second, the transistor-based generator is capable to adjust the pulse interval time t_o (Fig. 11.4) based on the EDM debris flushing condition. This enables the EDM motion control to adjust the gap distance between the electrode and workpiece. A longer ignition delay time means the electrode needs to travel a longer distance to initiate the spark and the gap distance between the electrode and workpiece is bigger. In practice, the motion control of an EDM machine electrode is controlled by adjusting the average cycle voltage to keep adequate gap distance. The average cycle voltage $\bar{u}$ is defined as (Fig. 11.4):

$$\bar{u} = \frac{u_e t_e + u_o t_d}{t_e + t_d + t_o}. \tag{11.4}$$

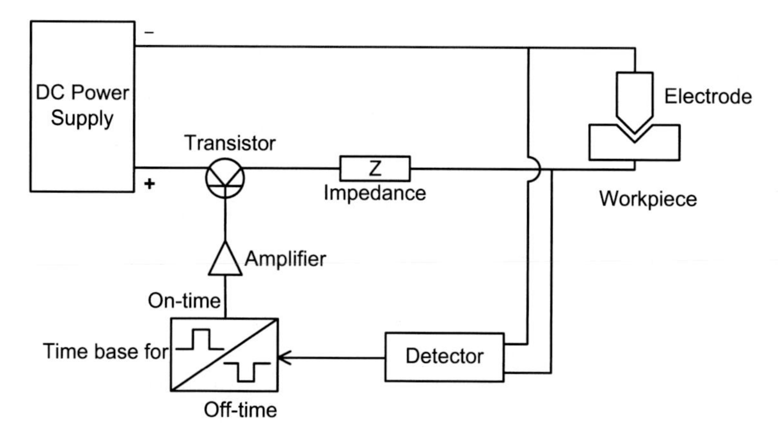

Fig. 11.6 Configuration of the transistor EDM generator

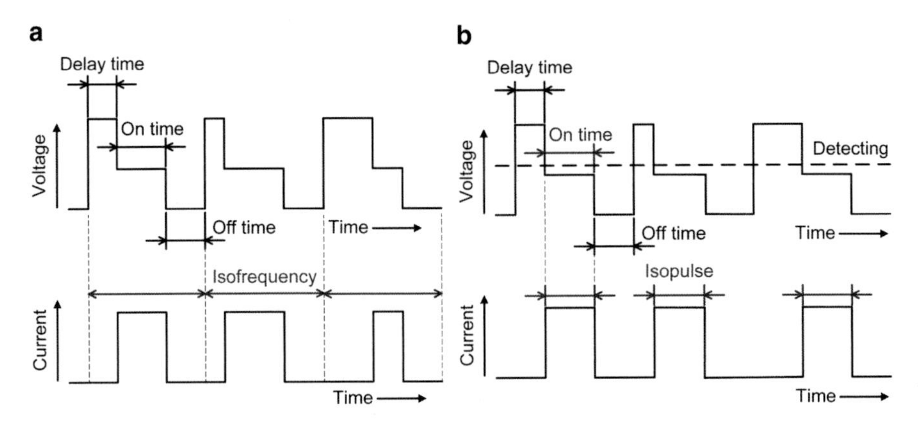

Fig. 11.7 The **a** isofrequency pulses by RC generator and **b** transistor generator enabled isopulse

Duty factor (DF) is defined as the ratio of time with discharge:

$$DF = t_e / (t_e + t_d + t_o). \tag{11.5}$$

$$E_d = u_e i_e t_e \tag{11.6}$$

A shortcoming of the transistor-based EDM generator is the limit to deliver very low discharge energy for finishing EDM to generate fine surface finish or micro features due to the cost and performance of high frequency MOSFET transistor. RC generators could be more cost-effective in finishing EDM applications. Some EDM machines incorporate both the transistor generator for roughing EDM and RC generator for finishing EDM.

Example

For an EDM machine with transistor-based generator, average cycle voltage $\bar{u} = 25\text{V}$, the discharge duration time $t_e = 100\,\mu\text{s}$, and pulse interval time $t_o = 40\,\mu\text{s}$. It is given that discharge current $i_e = 50\text{A}$, open voltage $u_o = 125\text{V}$, and discharge voltage $u_e = 20\text{V}$. For the ideal condition of isopulse EDM, calculate:

(a) Discharge delay time, t_d
(b) Cycle frequency, f_r
(c) DF
(d) Average energy of individual discharge, E_d
(e) Machining power
(f) MRR (assumed material removal volume 0.008 mm^3 per pulse)

Solution

(a) From Eq. (11.4), $\bar{u} = \dfrac{u_e t_e + u_o t_d}{t_e + t_d + t_o}$ the discharge delay time, $t_d = 15\,\mu s$
(b) Cycle frequency, $f_r = 1/(t_e + t_d + t_o) = 1/(100 + 15 + 40) = 6.45\text{kHz}$
(c) Duty factor, $DF = t_e/(t_e + t_d + t_o) = 0.645$
(d) Energy of individual discharge $E_d = u_e i_e t_e = 50 \times 20 \times 100 \times 10^{-6} = 0.1\text{J}$
(e) Machining power=0.1 (J) × 6.45 (kHz)=645 W
(f) MRR=0.008 (mm^3) × 6.45 (kHz)=51.6 mm^3/s

Example

A hardened steel is machined with EDM using either RC circuit or transistor-based pulse generator. Assume the same open voltage (u_o) of 200 V, the same cycle frequency and the same individual pulse discharge energy of 0.25 J, find:

(a) The values of R and C for the RC circuit with 15 μs discharge time (t_e) (given that $u_e = 0.73u_o$ and $t_e = 0.1t_c$).
(b) The average cycle voltage (u_s) and current (i_e) for transistor-based circuit with 20 μs discharge delay time, 0.75 DF, and 25 V discharge voltage (u_e).

Solution:

(a) For the RC circuit: At maximum removal rate, $u_e = 0.73u_o = 146\text{V}$

$E_d = \frac{1}{2}Cu_e^2 = 0.25(\text{J})\text{ then } C = 23\,\mu F$.

Because $t_e = 0.1t_c, t_e = 15\,\mu s, t_c = 150\,\mu s$. According to $u_c = u_o(1 - e^{-t/RC})$, $R = 4.98\Omega$.

Cycle frequency, $f_r = 1/(t_c + t_e) = 1/(150 + 15)(\mu s^{-1}) = 6060\text{Hz} = 6.06\text{ kHz}$.

(b) For transistor, the cycle frequency, f_r, is the same as RC circuit, then $t_e + t_d + t_o = 165\,\mu s$

$DF = t_e/(t_e + t_d + t_o) = t_e/165(\mu s) = 0.75, t_e = 123.8\,\mu s$.

Because $t_d = 20\,\mu s$, $t_o = 21.2\,\mu s$ and the individual discharge energy 0.25 J, $E_d = u_e i_e t_e = 0.25$ J.

Therefore, $0.25 = i_e \times 25 \times 123.8 \times 10^{-6}$ and $i_e = 80.8$ A.

The average cycle voltage $\bar{u} = \dfrac{u_e t_e + u_o t_d}{t_e + t_d + t_o} = \dfrac{25 \times 123.8 + 200 \times 20}{165} = 43.0\text{V}$.

11.6 EDM Surface Integrity

In the discharge at EDM, the heat flux is very high, could reach the order of 10^{17} W/m^2, during the discharge spark of a very short duration, typically in the range of 0.01–2000 μs. The peak temperature of the discharges can reach 10,000 °C. The

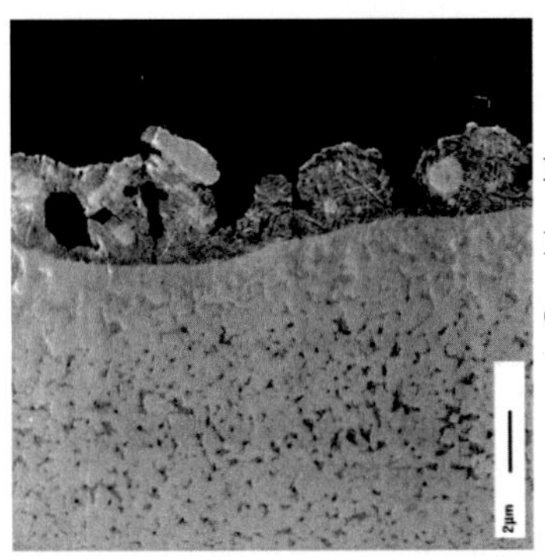

Fig. 11.8 Cross section of EDM surface of tungsten carbide in cobalt matrix

temperature is so high that it will melt both workpiece and electrode. As a result, a durable electrode is usually made of the high melting temperature materials, such as graphite or tungsten, to reduce the wear and improve life and EDM accuracy.

Due to the high temperature, metallurgical changes occur in the subsurface layers of the EDM workpiece. Commonly there are three zones that can be observed: (1) recast zone, (2) heat-affected zone, and (3) conversion zone. An example of the tungsten-carbide tool in cobalt matrix tool material EDM surface cross section is shown in Fig. 11.8. The surface of the workpiece is melted and quickly resolidified by the cooling action of the dielectric medium so that a thin recast layer is formed. Beneath the recast layer is the heat-affected zone, which is formed by the heating, cooling, and the diffusion of material from the melted recast layer. For example in Fig. 11.8, the cobalt binder material is melted and filled in the porous cavity in the heat-affected zone. Below the heat-affected zone is a conversion zone where modification in grain structure from the original structure occurs.

Surface integrity is important to the quality of the EDM parts. The thermal stresses in the melting process can weaken the grain boundary and reduce the fatigue strength on EDM surface. The generation of these metallurgically damaged surface layers motivates secondary finishing operations to be performed on parts produced by EDM before they are put into service. The removal of the recast layer and heat-affected zone has been found to increase the fatigue strength of the workpiece.

11.7 Conclusions

EDM utilizes a distinctly different material removal mode and is complementary to the traditional machining processes. EDM process has been widely used in many industrial applications and is the foundation of micromachining. For example, EDM is the critical and enabling process to generate the precision micro spray holes for clean combustion in internal combustion engines, the geometry and threads on precision dies and molds and the high aspect ratio cooling holes in aircraft engines. These are examples that are very difficult to accomplish using the traditional machining process but feasible with EDM.

Homework

1. In some produce of mold using EDM process, pulse discharge energy of 0.1 J is used with transistor-based generator. Open voltage of 150 V and discharge current of 25 A are required. Assume the discharge voltage of mold steel is 20 V. For the process with servo voltage of 25 V and DF 0.8, determine the:
   (a) Discharge time, t_e
   (b) Discharge delay time, t_d
   (c) Pulse interval, t_o
   (d) Cycle frequency, f_r
   (e) EDM machining power
2. Microholes on hardened steel will be machined with EDM using RC circuit, $R = 20\ \Omega$ and $C = 5\ \mu F$. The energy of individual discharge is controlled at 0.025 J for maximum removal rate, to calculate:
   (a) What voltage should be set for power supply?
   (b) Charging time, t_c
   (c) Cycle frequency, f_r
   (d) Peak current, i_p
   (e) If the dielectric medium has dielectric strength of 8 V/μm, estimate the gap distance
3. A hardened steel will be machined with EDM, we consider pulse generate circuit with RC circuit or transistor circuit. For the conditions of the same power supply voltage as 200 V and individual pulse discharge energy as 0.25 J.
   (a) If discharge duration time of both are set at 15 μs. Please design the RC circuit and transistor circuit separately to obtain same cycle frequency. Assume the discharge voltage of hardened steel is 25 V, its ideal delay time is 20 μs (Hint: RC circuit is required to be performed at maximum removal rate condition, $\ln 0.27 = -1.3093$).
   (b) If MRR is proportional to the 0.5 power of discharge current, please compare the efficiency of transistor circuit to RC circuit.

Electrochemical Machining, Chemical Machining, and Chemical Mechanical Polishing Processes

12

Electrochemical machining (ECM) is a machining process which removes material through anodic electrochemical dissolution. ECM can create complicated forms in high-strength, temperature-resistant difficult-to-machine materials because of the minimal tool (electrode) wear. For example, ECM is commonly used in the manufacturing of turbine blades made of Ni-based high-temperature alloy. Chemical machining (ChM), also known as etching or chemical etching, is the controlled chemical dissolution of work material for machining. It is a critical process to manufacture large components as well as micro features in integrated circuits (IC). Chemical mechanical polishing (CMP), also known as chemical mechanical planarization, utilizes the combination of chemical and mechanical effects to polish and flatten surfaces. It is a critical process for IC manufacturing. These three nontraditional machining processes related to chemical actions are presented in this chapter.

12.1 Electrochemical Machining

Electrochemical machining (ECM) utilizes the electrolytic process to dissolve the work-material. The workpiece is the anode and the tool electrode is the cathode. Using high-density, low-voltage direct current (DC), the work-material in anode is dissolved into metallic ions and removed atom by atom.

Electrolysis is a chemical process in which an electric current is passed between two ferrous conductors dipped in a liquid solution. When an electrical potential difference is applied across the two conductors, or electrodes, the dissolution of iron takes place at the anode:

$$Fe \rightarrow Fe^{2+} + 2e^{-}.$$

At the cathode, hydrogen gas and hydroxyl ions are generated:

S. Y. Liang, A. J. Shih, *Analysis of Machining and Machine Tools*,
DOI 10.1007/978-1-4899-7645-1_12

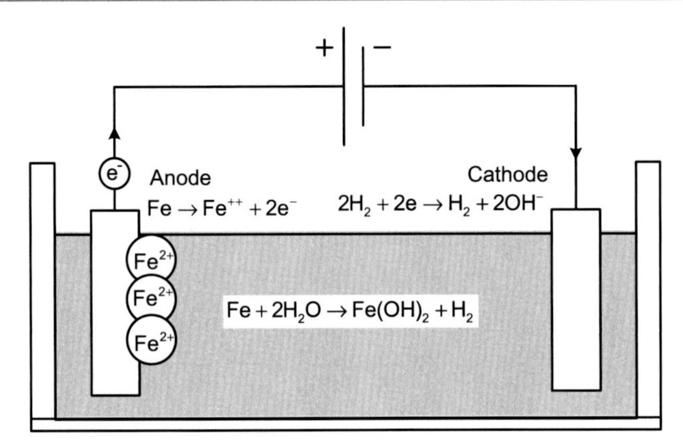

Fig. 12.1 The electrolysis of iron

$$2H_2O + 2e^- \rightarrow H_2 + 2OH^-.$$

The outcome is that the metal ions combine with the hydroxyl ions to precipitate out as iron hydroxide:

$$Fe + 2H_2O \rightarrow Fe(OH)_2 + H_2.$$

Electrolysis involves the dissolution of iron from the anode and the generation of hydrogen at the cathode as shown in Fig. 12.1. No other action takes place at the electrodes. Since only hydrogen gas is evolved at the cathode, the shape of the cathode remains the same during this process.

The fundamental concept of ECM, the electrolysis, can be described by Faraday's law of electrolysis which was first published in 1834. In 1929, a Russian researcher, Gussef, filed a patent to use ECM process for machining metal. It was not until 1959 that the first commercial ECM machine was established by Anocut Engineering Company. After that, ECM advanced throughout the 60s and 70s and gained popularity, especially in aerospace engine and power-generation gas turbine industry.

In ECM, the workpiece and tool (in the form of a die) are made the anode and cathode, respectively. The electrical potential difference, usually of about 10–20 V, is applied across them. The use of electrolyte (usually aqueous NaCl, $NaNO_3$, NaOH, KCl, and other solutions) ensures that the cathode shape remains unchanged during electrolysis. The electrolyte is also pumped through the gap between the electrodes to remove the accumulated of metallic and gaseous products of the electrolysis. If such accumulation is left uncontrolled, eventually a short circuit would occur between the electrodes.

As machining proceeds, the gap between the electrodes will gradually evolve into a steady-state value, and the shape of the cathode will be reproduced on the anode. The basic configuration of the process is shown in Fig. 12.2.

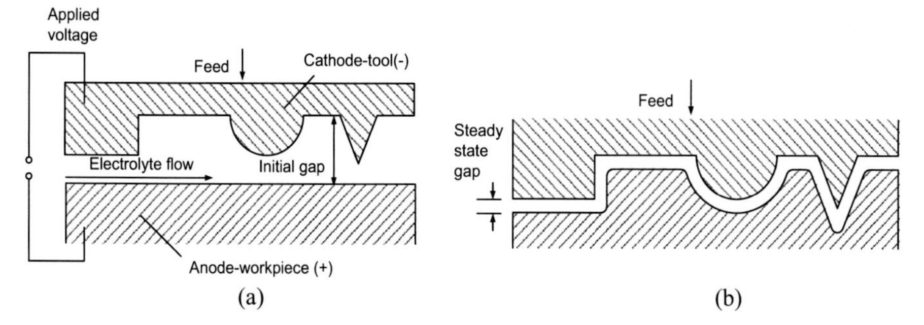

Fig. 12.2 Overview of the ECM process showing the tool–workpiece configuration **a** initially, and **b** at final stage

Faraday's law states that the rate of dissolution ($\dot{v}\rho$) depends only upon the atomic weight of the electrode (w), valency of electrode material (z), and the current applied (I)

$$\dot{v}\rho = \frac{wI}{zF}, \tag{12.1}$$

where v is the volume removed from the anode, ρ is the density of the anode metal, and F is Faraday's constant (96,500 C). The electrochemical equivalent (k_e) of the workpiece material can be defined by

$$k_e = \frac{w}{zF}. \tag{12.2}$$

Note that the dissolution rate is not affected by the hardness or other characteristics of the electrode metal. It follows from Ohm's law that the current I is proportional to the applied voltage (V).

$$I = \frac{V}{R}, \tag{12.3}$$

where R is the resistance of the conductor, which is the electrolyte in the case of ECM. This resistance is in turn proportional to the gap size between the cathode and the anode (h) and inversely proportional to the cross-sectional area of the electrodes (A)

$$R = \frac{h}{kA}, \tag{12.4}$$

where k is the specific conductivity as a property of the electrolyte. Therefore, the required current density (J)—current per electrode area—is

$$J = \frac{I}{A} = \frac{kV}{h}. \tag{12.5}$$

Combining Eqs. (12.1)–(12.5), the volumetric removal rate ($\dot{v}$) can be calculated by

$$\dot{v} = \frac{VkAw}{zh\rho F} = \frac{VkAk_e}{h\rho}. \tag{12.6}$$

Example

Iron has a density (ρ) of 7800 kg/m^3, valency (z) of 2, and an atomic weight (w) of 0.028 kg. When it is electrochemically machined with an electrolyte of specific conductivity (k) of 0.2 Ω^{-1} cm^{-1}, electrode of 2.5 cm^2 area (A), an electrode gap (h) of 0.0062 cm, and an applied potential (V) of 10 V, what is the expected volumetric removal rate? What is the current density under this condition?

Solution:

$$\dot{v} = \frac{VkAw}{zh\rho F} = \frac{10(0.2)(2.5)(0.028)}{2(0.0062)7800(96{,}500)} = 1.5 \times 10^{-8}\ \text{m}^3/\text{s}.$$

$$J = \frac{kV}{h} = \frac{0.2(10)}{0.0062} = 322\frac{\text{A}}{\text{cm}^2}.$$

Example

A cobalt (Co) plate of 1000 mm^2 surface area is machined by ECM to reduce its thickness by 2 mm. The atomic weight of Co is 58.93 and its density is 0.00892 g/mm^3 (8920 kg/m^3). The ECM machine provides constant voltage of 10 V and uses NaCl at 2 M/L concentration which has electrical conductivity of 0.020 (S mm^{-1}). The gap distance is controlled to be 0.1 mm.

a. What is the total working current?
b. What is the machining time required?

Solution:

a. $$R = \frac{h}{kA} = \frac{0.1 \times 10^{-3}}{20 \times 1000 \times 10^{-6}} = 0.005\ \Omega$$

$$I = \frac{V}{R} = \frac{10}{0.005} = 2000\ \text{A}.$$

b. $Co \rightarrow Co^{2+} + 2e^-$.

$$k_e = \frac{w}{zF} = \frac{58.93}{2 \times 96,500} = 3.05 \times 10^{-4}.$$

Volume to be removed is $2 \times 1000 = 2000\ \text{mm}^3$

$$\dot{v} = \frac{VkAk_e}{h\rho} = \frac{10(0.020)(1000)3.05 \times 10^{-4}}{(0.1)(0.00892)} = 68.39\ \text{mm}^3/\text{s}.$$

$$t = \frac{\text{Volume}}{\dot{v}} = \frac{2000}{68.39} = 29.2\ \text{s}.$$

The ECM process involves the application of a fixed voltage between electrodes while the cathode is driven toward the workpiece at a constant velocity. Due to the dissolution of workpiece material coupled with the relative motion between the die and the workpiece, the gap size can vary with time and this dynamic aspect is governed by

$$\dot{h} + f = \frac{wI}{\rho FA} \Rightarrow \dot{h} = \frac{wJ}{\rho F} - f, \tag{12.7}$$

where f is the cathode feed rate as shown in Fig. 12.3. Substituting Eq. (12.5) in Eq. (12.7) gives

$$\frac{\mathrm{d}h}{\mathrm{d}t} = \frac{wkV}{\rho Fh} - f. \tag{12.8}$$

For known initial condition of $h(0)$, the above can have two practical cases:

Case I: No tool movement, $f=0$; Eq. (12.8) has the solution for gap $h(t)$

$$h^2(t) = h^2(0) + \frac{2wkVt}{\rho F}, \tag{12.9}$$

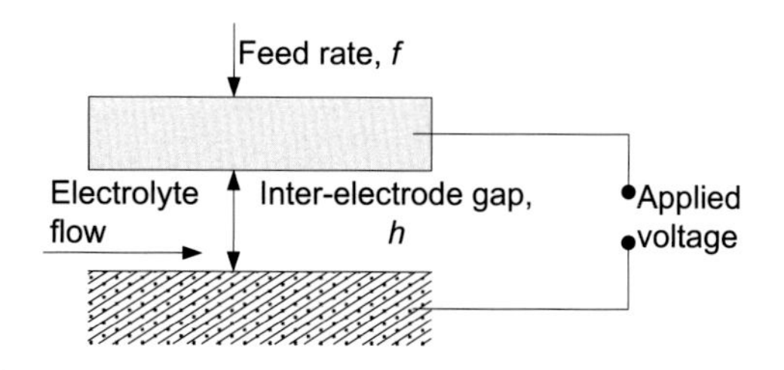

Fig. 12.3 Electrodes in ECM

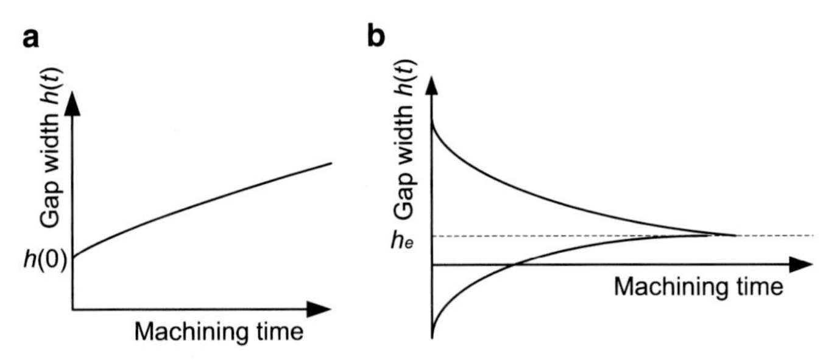

Fig. 12.4 **a** Zero feed rate scenario and **b** constant feed rate cases

as plotted in Fig. 12.4a. The gap width increases indefinitely with the square root of the machining time. This condition is often used in ECM deburring when surface irregularities are removed from components in a few seconds.

Case II: Constant feed rate (f): Eq. (12.8) then has the solution of

$$h(t) = h(0) - tf + h_e \ln \frac{h(0) - h_e}{h(t) - h_e}, \tag{12.10}$$

where $h_e = \dfrac{wkV}{F\rho f}$ is the steady-state value for $h(t)$.

Two solutions of Eq. (12.10) are given in Fig. 12.4b. Note that these values are independent of the initial gap width $h(0)$; the process thus can be used to reproduce an irregular shape of the cathode on the workpiece.

Example 12.2

Considering the previous example under a feed rate of 0.02 mm/s, what is the time it takes to achieve 90 % of the steady-state value of gap width?

Solution:

$$h_e = \frac{wkV}{F\rho f} = \frac{(0.028)(0.2)(10)}{(96{,}500)(7800)(0.02)} = 3.7\times10^{-2}\,\text{cm} = 0.037\ \text{cm}.$$

$$h(0) = 0.0062\ \text{cm},\ h(t) = 0.9h_e = (0.9)(0.037)\ \text{cm}.$$

Based on Eq. (12.10),

$$(0.9)(0.037) = 0.0062 - (0.002)t + (0.037)\ln\frac{0.0062 - 0.037}{(0.9)(0.037) - 0.037}.$$

Then, the solved time is $t = 25.7$ s.

12.2 Chemical Machining

Chemical machining (ChM) is a nontraditional machining process which removes workpiece material by controlling chemical dissolution with a strong acidic or alkaline chemical reagent. Selective machining can be done by coating a maskant material on the workpiece surface to protect areas from which the metal is not to be removed. ChM is used to machine high-strength materials, large and thin workpiece, such as wings of a jet plane, and micro features. ChM can be traced back to as early as 2300 BC when Egyptians used citric acid to machine copper. Before the twentieth century, ChM was mainly used in decorative etching and printing. The semiconductor and microelectromechanical systems (MEMS) industry has transformed ChM into a key manufacturing process that greatly influences the yield and quality of the IC and MEMS products.

There are five major steps of ChM,

1. Workpiece preparation: A clean workpiece surface is required to allow for good adhesion of masking material. Dust oil, and rust have to be removed by a cleaning process. This process can be done mechanically or chemically.
2. Applying of masking material: The purpose of this step is to selectively protect the workpiece from chemical etching. Methods of applying masking material include printing, flowing, dipping, spraying, and spin coating. Polymer and rubber are two common masking materials. The selection of masking material depends on both the etchant and the workpiece material. The masking material needs to be inert to the etchant and able to adhere to the workpiece surface. A good masking material also has to be tough enough to withstand handling, patterning, and removal. The resolution and the accuracy of ChM are dependent on masking material. Photoresist has the best resolution. The cut-and-peel maskant provides poor resolution but allows for a deeper etching.
3. Patterning masking material: Once the masking material has been applied, the desired pattern of masking material will be scribed. This pattern only exposes the areas that will receive chemical machining. Patterning method is masking material dependent. For example, a laser cutter can be used to pattern cut-and-peel maskant, and a photomask and light are used to pattern photoresist. This process is also known as photolithography and is widely used in semiconductor manufacturing.
4. Etching: The chemical reagent used to etch and dissolve the workpiece is called etchant. The workpiece material is removed by immersing into selected etchant and those areas not covered by masking material will be removed. This process is usually controlled by time with predetermined material removal rate (MRR), which will be introduced later. Etchant is workpiece material dependent. Two typical etchants are caustic alkaline and acidic solutions. Besides workpiece material, it is important to select a proper etchant based on type of masking material, etching depth, required surface roughness, desired MRR, cost, and environmental and safety issues. Table 12.1 lists some common etched materials and etchants.

Table 12.1 Common etched materials and their etchants

Etched material	Etchant
Aluminum alloys	Sodium hydroxide
Magnesium alloys	Sulfuric acid
Steel/nickel alloys	Hydrochloric, nitric, sulfuric, phosphoric acids
Titanium alloys	Hydrofluoric acid with nitric or chromic acid
Copper alloys	Ferric chloride

Traditionally, ChM is done by immersing the workpiece into etchant solution. This process is named "wet etching." "Dry etching" is also available, which removes material by bombardment of ions. The ion source can be plasma of reactive gas such as oxygen and fluorine. Dry etching produces a part with better accuracy.

5. Cleaning: After etching, the workpiece is rinsed to clean out the etchant. Then, the masking material is removed by hand-stripping or by immersing into a suitable demasking solution.

The advantages and disadvantages of ChM are summarized in Table 12.2.

The MRR in ChM is usually expressed as chemical etch rate, v_{ch} (in the unit of mm/s or similar). With known workpiece material and etchant, v_{ch} can be estimated by:

$$v_{ch} = \frac{10^{-6} E_{ch} D_{ch} M_w}{\rho_w N_{ch} \delta}, \tag{12.11}$$

where

E_{ch} is the bulk concentration of chemical etching species (mole/L)
D_{ch} is the diffusion coefficient of etchant in solution (mm^2/s)
M_w is the molecular weight of workpiece

Table 12.2 Advantages and disadvantages of ChM

Advantages	Disadvantages
Easy weight reduction	Difficult to get sharp corner
Easy and quick design changes	Difficult to chemically machine thick material (the maximum thickness is about 10 mm)
Low tooling costs	Etchants are dangerous for workers
No burr formation	Etchant disposals are very expensive and not environmentally friendly
Good surface quality	
No effect of workpiece materials properties such as hardness	
Low capital cost of equipment (not applied for photolithography)	

N_{ch} is the number of etchant molecules to dissolve one workpiece molecule
ρ_w is the density of workpiece (g/mm^3)
δ is the diffusion layer thickness (mm)

ChM process is generally controlled by machining time. With the calculated v_{ch} and desired removal thickness, the required machining time can be determined.

Example
In a semiconductor manufacturing process, hydrogen fluoride (HF) is used to etch a silicon dioxide layer with 50-μm depth. Given the chemical reaction of oxide wet etching is:

$$SiO_2 + 6HF \rightarrow H_2SiF_6 + 2H_2O.$$

The concentration of HF used in this process, E_{ch}, is 7 mole/L.

For SiO_2, its molecular weight, M_w, is 60 g/mole and density, ρ_w, is 2.634×10^{-3} g/mm^3.

The diffusion layer thickness, δ, is 0.4 mm.

The diffusion coefficient, D_{ch}, is 0.01 mm^2/s.

Estimate the etching time of this process, assuming there is no undercut.

Solution:

Given $N_{ch} = 6$, $E_{ch} = 7$ mol/L, $M_w = 60$ g/mole, $\rho_w = 2.643 \times 10^{-3}$ g/mm^3, $\delta = 0.4$ mm, and $D_{ch} = 0.01$ mm^2/s.

$$v_{ch} = \frac{10^{-6} E_{ch} D_{ch} M_w}{\rho_w N_{ch} \delta} = \frac{10^{-6} \times 7 \times 0.01 \times 60}{2.643 \times 10^{-3} \times 6 \times 0.4} = 6.62 \times 10^{-4} \text{ mm/s}$$

$$t = \frac{\text{Depth}}{v_{ch}} = \frac{50 \times 10^{-3}}{6.62 \times 10^{-4}} = 75.5 \text{ s.}$$

12.3 Chemical Mechanical Polishing

Chemical mechanical polishing (CMP) combines both chemical and mechanical effects to polish and flatten the surfaces. With the decreasing feature size in an IC manufacturing, CMP is a key process to allow for an accurate and precise production of IC. Currently, CMP is the only technique that can reach both local and global planarity on the wafer surface.

The schematic of CMP process is shown in Fig. 12.5. The wafer surface to be polished is placed against a pad, which is fixed on the platen. The pad carries slurry and provides support and polishing action against the wafer surface. The slurry is

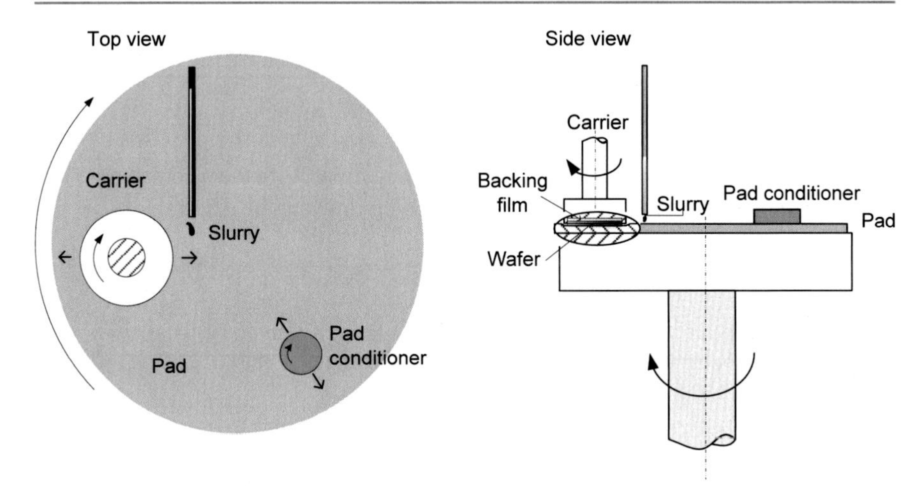

Fig. 12.5 Schematic of CMP

dispensed from a tube, and as the platen rotates the slurry is transported between the pad and the wafer. The slurry may react with the wafer and form an active layer that is easier to be removed, as shown in Fig. 12.6. This is the chemical reaction in CMP. There are abrasive particles in the slurry causing mechanical damage to the wafer surface, loosening the material for an enhanced chemical attack, or fracturing off the pieces of surface into a slurry where they dissolve or are swept away. The process is tailored to provide an enhanced material removal rate from high points on the surface (compared to low points), thus affecting the planarization. Note the chemistry alone will not achieve planarization, because most of the chemical reactions are isotropic. The mechanical grinding alone, theoretically, may achieve the desired planarization but is not desirable because of extensive associated damage to the material surfaces, like macro scratch.

It can be concluded that there are three main components in CMP.

1. Wafer: The wafer surface is polished. P
2. Pad: The key media enabling the transfer of mechanical forces to the surface being polished.
3. Slurry: The slurry provides both chemical and mechanical effects.

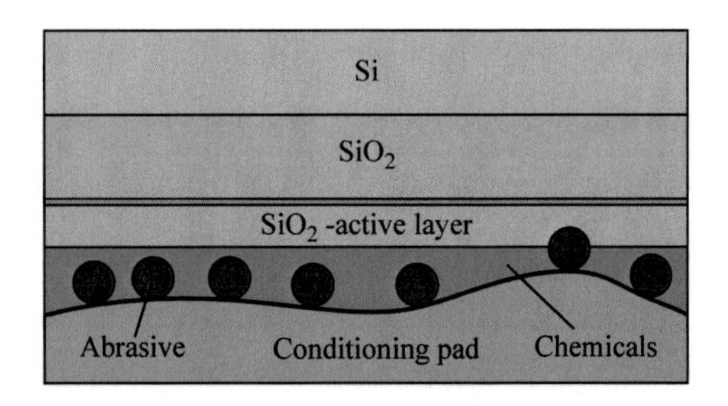

Fig. 12.6 Material removal in CMP

There are numerous indices to evaluate the performance of CMP. The within-wafer-non-uniformity (WIWNU), wafer-to-wafer-non-uniformity (WTWNU), MRR, and defect counts are the common ones used in industry.

The Preston's equation on $MRR = K_p P^a V^b$ is the first and most commonly used form for the CMP MRR. It is an empirical equation, which states that the average MRR is linearly proportional to the product of the applied down force (pressure) and the relative velocity between wafer and pad surface. However, in practical application, the relationship between MRR and down force is usually nonlinear. So, later studies proposed some nonlinear models, such as $MRR = K_p (PV)^{1/2}$, $MRR = K_p P^{2/3} V$, and other models considering abrasive particle sizes, pad surface roughness, and other polishing conditions.

Homework

1. In the ECM operation given in the figure, determine the minimum required time such that when the base plane of the cathode comes within 1 mm proximity of the workpiece, the tip of the cathode has already achieved a steady-state gap width relative to the workpiece. For this problem, $w = 0.028$ kg, $\rho = 7800$ kg/m^3, the specific conductivity is 0.3 Ω^{-1} cm^{-1}, the initial electrode gap between the base plane and workpiece is 2 mm, and the applied potential is 15 V. Assume constant feed rate for this problem.

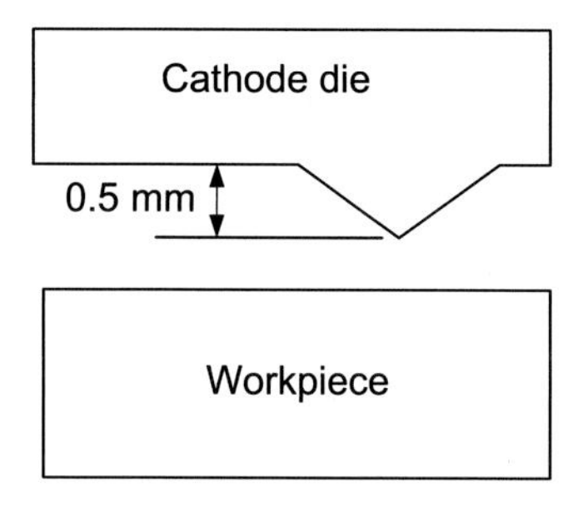

2. A weight-reducing process will be conducted to machine a titanium (atomic weight 47.9 g, density 0.00452 g/mm^3, and valency 3) workpiece with surface area of 2500 mm^2 using an ECM machine. This ECM machine provides a voltage of 20 V and can afford current up to 5000 A. The electrolyte used in this machine has electrical conductivity of 20 (S m^{-1}).
   a. What is the smallest possible gap distance?
   b. How long does it take to reduce weight by 50 g?

13 Laser and Electron Beam Machining

This chapter introduces two nontraditional machining processes: laser machining and electron beam machining (EBM). These two machining processes use laser and electron beams as energy sources to introduce thermal energy to remove work material by melting and/or vaporization. The chemical reaction between the photon/electron and workpiece is another material removal mechanism using laser. When performing precision micromachining, it is important to consider the chemical reaction to achieve the accurate dimension and desired surface integrity. Applications of laser beam machining (LBM) and EBM processes are broad, from large-scale cutting to microscale drilling of work materials ranging from hard metals to soft polymers.

13.1 Laser Beam Machining

13.1.1 History and Overview

Laser, first invented by Gordon Gould in 1957, is the acronym of light amplification by stimulated emission of radiation. Laser is a coherent, monochromatic, and collimated light that can be focused to a small area to produce high power density for machining. The temporal and spatial coherence is required for the laser beam to obtain useful power density. Temporal coherence means the laser has highly collimated wavelength. Spatial coherence means the laser is focused to a narrow beam with low divergence. The useful power density is achieved by focusing, thus attaining adequately high power density to process work materials.

LBM is used to cut and machine both hard and soft materials, irrespective of its physical and mechanical properties. CO_2 laser is a traditional and common energy source for LBM of a variety of materials. For high-thermal-conductivity materials, such as Al, Cu, Ag, and Au, LBM is especially problematic, as these metals conduct the heat away easily and have the tendency to reflect the light. The yttrium aluminum garnet (YAG) laser has been used to cut these metals after treating their

S. Y. Liang, A. J. Shih, *Analysis of Machining and Machine Tools*,
DOI 10.1007/978-1-4899-7645-1_13

surfaces by oxidation or increasing their roughness. YAG laser sometimes is superior to CO_2 laser because it emits a shorter wavelength. High capital and operating cost and low energy efficiency ($\eta = 1\,\%$) often prevent LBM from being competitive with other nontraditional machining techniques.

The history of laser and LBM is briefly summarized in Table 13.1.

As mentioned before, a laser beam has highly collimated wavelength and is monochromatic. Laser with different wavelengths can be excited from different lasing medium, which will be introduced later. Figure 13.1 illustrates common wavelengths that are used in LBM. The wavelength covers the range of both visible and invisible wavelength.

13.1.2 Classification of LBM

The laser beam can be output with two modes: continuous wave (CW) mode and pulsed (P) mode. Generally, CW lasers are used for processes like welding, soldering, and surface hardening that require an uninterrupted supply of energy for melting and phase transformation. Controlled P mode is desirable for cutting, drilling, marking, and so on, striving for less heat distortion and a minimum heat-affected zone (HAZ). P mode can be classified by the pulse duration into micro-, nano-, pico-, femto-, and atto- second laser in the 10^{-6}, 10^{-9}, 10^{-12}, 10^{-15}, and 10^{-18} s range, respectively.

Table 13.1 History of laser and laser beam machining (LBM)

Year	Event
1957	Gordon Gould invented the modern laser
1960	Ruby laser, HeNe gas laser
1962	Yttrium aluminum garnet (YAG) laser
1963	Fiber laser
1964	CO_2 laser and neodymium-doped yttrium aluminum garnet (Nd:YAG) laser
1965	Solid state laser, chemical laser, LBM for microhole
1966	Dye laser
1968	Laser applied in medicine (eye surgery)
1970s	Industrial applications in welding
1975	KrF, XeF, XeCl excimer laser
1982	Short pulse ultraviolet (UV) laser for polymer
1986	Excimer laser for micromachining

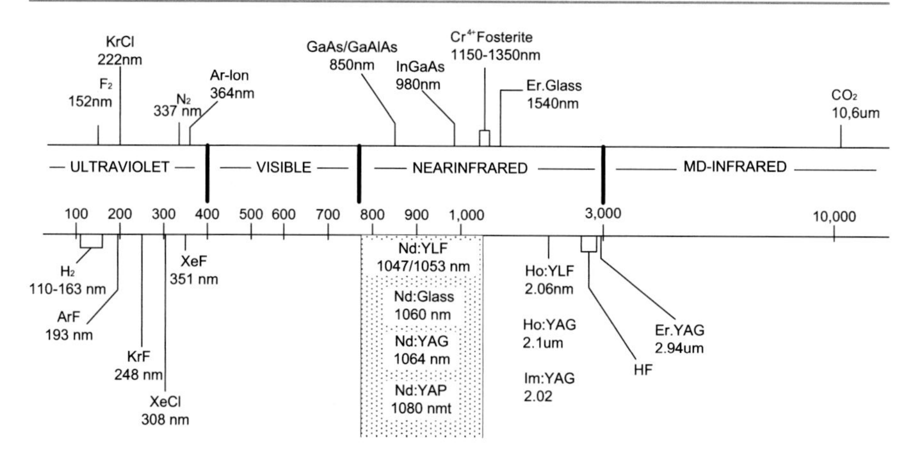

Fig. 13.1 Wavelength of laser for LBM. *YAG* yttrium aluminum garnet, *YAP* yttrium aluminum perovskite, *YLF* yttrium lithium fluoride, *HF* hydrogen fluoride

Laser can also be classified by its lasing medium, which is the medium that allows for the emission of coherent radiation. Laser is named after its lasing medium. Three major categories of lasing mediums are solid, gas, and semiconductor. Other lasing mediums include the chemical, organic dye, metal vapor, and Raman scattering. The lasing medium determines the wavelength of laser. Table 13.2 summarizes some typical lasing medium and its corresponding laser wavelength. Solid laser is usually used in pulse model and has low-pulsing frequency (1–2 pulses/s). Instead of lasing, optical fiber in the form of thin, long fiber can be the medium for the fiber

Table 13.2 Lasing mediums of laser

Category	Lasing medium		Wavelength (λ) (nm)
Solid	Ruby		694.3
	Nd:YAG ($Y_3Al_5O_{12}$)		1064
	Nd-glass (glass rod doped with 2–6% Neodymium)		1054–1062
Gas	CO_2		10,600
	Excimer	F_2	157
		ArF	193
		KrCl	222
		KrF	248
		XeCl	308
		XeF	351
Semiconductor	GaAs (gallium arsenide)		840
	GaN (gallium nitride)		405
	InGaAs (indium gallium arsenide)		980

Nd:YAG neodymium: yttrium aluminum garnet

laser. The fiber laser is compact and can have high power and be directed directly to the optical focus element.

CO_2 laser is a common laser source for high-power laser machining. CO_2 laser uses CO_2 gas as the lasing medium and is characterized by their long wavelength of 10,600 nm, which is in the infrared (IR) range. A mixture of gases is also used (CO_2:N_2:He=0.8:1:7). In this case, helium acts as a coolant for the gas cavity. CO_2 lasers yield the highest depth-to-diameter ratio in most metals using gas-jet assistance. CO_2 lasers are bulky but efficient. The ratio of output power to pumping power can be as high as 20 %. There are two types of CO_2 lasers. One is the axial-flow CO_2 laser. This laser can operate in both CW and P modes. Another is the transverse-flow CO_2 laser, which operates only in the CW mode and is used when high-power outputs are required. The CW CO_2 ranges from mW to GW, which leads to broad applications from industrial engraving, machining, and welding to medical procedures such as skin resurfacing. Water, the major composition of the tissue, absorbs the CO_2 laser very well and makes it suitable for surgical cutting and joining applications.

Nd:YAG is a common lasing medium used in industry and medicine. It is a single crystal of YAG doped with 1 % neodymium and can operate in either CW or P mode. Nd:YAG laser has relatively high efficiency and pulsing frequency. For manufacturing, Nd:YAG laser is common for cutting, welding, engraving, and marking of metals and plastics with power typically ranging from 1 to 5 kW. In medicine, Nd:YAG laser is utilized in cutting of soft tissue (hair, skin, prostate, etc.), ophthalmology (e.g., posterior capsulotomy), and laser-induced thermotherapy.

Excimer is an abbreviation of "excited dimmer." The beam is generated due to fast electric discharging in a mixture of high-pressure dual gas, composed of one from the halogen gas group (F, H, Cl) and another from the rare gas group (Kr, Ar, Xe). The wavelength of the excimer laser attains a value from 157 to 351 nm, which is in the UV region of the spectrum. The wavelength of excimer laser depends on the dual gas combination. Excimer lasers have low wavelengths and high photon energy, so they can remove or ablate material photochemically (instead of the thermal melting or vaporization) and have remarkable precision and efficiency in the microscale machining applications for plastics and biological tissues. It is common in semiconductor high-resolution photolithography (e.g., deep UV of KrF and ArF), ophthalmology (e.g., laser-assisted in situ keratomileusis, LASIK), and removal of plaque inside the blood vessels in peripheral artery (e.g., laser athrectomy using XeCl).

Table 13.3 summarizes main characteristics of lasers that are commonly used in the industry.

13.1.3 LBM Material Removal Mechanisms and Processes

Material removal mechanisms of LBM can be classified by laser's interaction with the workpiece and dependent on the wavelength of the laser beam. Laser beam interacts with the workpiece material either thermally or chemically. In LBM the workpiece material interacts with the laser beam through several effects, including

Table 13.3 Main characteristics of industrial lasers

Laser type	Mode	Beam characteristics			Comments
		W_{av} (W)	f_r (p/s)	d_f (μm)	
CO_2	P	250–5000	400	75	High efficiency, bulky
	CW	100–2000	–	75	
	CW (transverse)	2500–15,000	–	75	
Nd:YAG	P	100–500	1–10,000	13	Compact, economical
	CW	10–800	–	13	
Nd:glass	P	1–2	0.2	25	Often uneconomical
Excimer	P	~100	10–500		Micromachining, plastic, ceramic

W_{av} = average power, f_r = pulse/s, and d_f = focus diameter
CW continuous wave, *P* pulsed

reflection, absorption, and conduction of absorbed heat that is followed by melting and evaporation. The unreflected light is absorbed, thus heating the surface of the workpiece. With sufficient heat, the workpiece starts to melt and evaporate. The physics of laser machining is complex due to scattering and reflection of laser light as well as the interference of the vapor and light on laser–surface interactions. Additionally, heat diffusion into the bulk material causes phase change, melting, and/or vaporization. Depending on the power density and time of beam interaction, the mechanism progresses from heat absorption and conduction to melting and then vaporization. High-intensity laser beams may not be effective since they form a plasma plume at or near the surface of the material with a consequent reduction in the process efficiency due to absorption and scattering losses.

In thermal-based LBM, the work material removal occurs by melting and vaporization of the work material. This mechanism is used mainly in applications such as cutting, drilling, welding, and surface hardening. The material removal rate (MRR) depends on the workpiece (e.g., thickness, surface roughness, and orientation) and material properties, such as thermal conductivity, specific heat, latent heat for melting and vaporization, and surface reflectivity. When power density reaches a threshold value (typically more than 10^6 W/cm^2), evaporation of material on the surface will become a high-density plasma, resulting in reduced absorption of the laser beam. This is called the plasma shielding effect. The ultrashort pulse, such as femtosecond, laser could potentially avoid this plasma shielding effect using the ablation (not thermal) material removal mechanism.

In chemical-based LBM, the material is removed by the dissociation and breaking of the chemical bond between the material molecules, when its bond energy is below the photon energy of the beam. The photon energy of the beam is inversely proportional to its wavelength. The shorter the wavelength is, the larger the photon energy will be. There are three steps during chemical-based LBM. First, the ultrashort wave photons are absorbed into the surface. Second, the chemical bond between molecules is broken. And finally, the reaction products escape as gas and small particle ashes. Fluorine excimer laser, as an example, is a laser of ultrashort wavelength ($\lambda = 157$ nm); thus, it possesses a high photon energy of 7.43 eV. This excimer laser is capable of machining plastic and teflon chemically, as its photon energy is greater than the chemical bond energy, which ranges from 1.8 to 7 eV ($1\ eV = 1.6 \times 10^{-19}$ J) for most of the plastics. On the other hand, CO_2 laser is an IR laser beam of long wavelength ($\lambda = 10{,}600$ nm) and with low photon power of 0.12 eV. Thus, the CO_2 laser is not capable of machining plastics photochemically, but photothermally.

Three key elements of the LBM equipment are: (1) the lasing material and its wavelength, (2) the pumping energy source required to excite the atoms of the lasing material to a higher energy level, and (3) the mirror system to direct the laser beam. One of these mirrors is fully reflective, while the other one is partially transparent to provide the laser output (output mirror). It allows the radiant beam to either pass through or bounce back and forth repeatedly through the lasing material.

A common industrial application of LBM is drilling. CO_2 and Nd:YAG lasers are commonly used in both the CW and P mode. LBM drilling can be done by either percussion or trepanning. Percussion drilling is often performed using the pulsed Nd:YAG lasers because of their higher pulse energy. Trepanning is used for large holes (>1 mm diameter) or small precision holes, such as those for diesel engine fuel injectors. The LBM cutting can be accomplished by either CW or P mode. CO_2 and Nd:YAG lasers are two common lasers used for cutting. When using CO_2 laser, CW mode is used for thicker metal sections, while P mode is used for thinner metal sections. CO_2 laser is commonly used because of its fast machining rate.

Gas is often used to assist LBM to either increase the machining rate or improve the surface finish. The assisting gas is supplied by a coaxial nozzle with continuous jet of air, O_2, or inert gases (e.g., N_2, Ar, or He). In the case of O_2, it promotes exothermic reaction and thus enhances the process efficiency by more energy needed for cutting. Inert gas is used to prevent plastics and other organic materials from charring or to promote an oxide-free surface of high-quality cut. The oxide-free edges can improve the weldability and part quality. The pressurized gases (with pressures of about 1–3 atm for O_2 and 2–6 atm for inert gas) can expel molten metal and vapors from the LBM zone. Selection of the gas depends upon the type of work material, its thickness, and type of cut.

LBM is a key micromachining process using two configurations: the direct writing and mask projection. For direct writing, the laser beam is moved by stages to "write" the desired profile instead of using a mask. It can achieve high resolution but has a low machining rate. It usually uses low-power laser and has precision focus. For mask projection, the laser beam goes through the mask and the projection lens to machine a miniature profile of the mask in the workpiece. The quality of machined edges relies on the resolution of the optical projection system. This configuration is useful for mass production, particularly in the semiconductor industry. There is also a configuration of contact mask writing in which the mask is in contact with the workpiece directly. The profile of the mask will be copied into the workpiece. It is simple to use and fast and can achieve the resolution of 10 μm or smaller. The disadvantages of contact mask writing are damages to the workpiece caused by the contact with the mask, especially for edges of the workpiece and to the mask which is exposed to the same energy level as the workpiece.

13.1.4 Power Density in LBM

A model to study the thermal-based LBM percussion drilling is shown in Fig. 13.2. The input and focus of LBM are converted to thermal energy to vaporize the workpiece material. The size of the spot diameter d_s is determined by

$$d_s = F_1 \theta, \tag{13.1}$$

where F_l is the focal length of lens and θ is the beam divergence angle (rad). The area of the laser beam at focal point, A_s, is

$$A_s = \frac{\pi}{4}(F_1 \theta)^2. \tag{13.2}$$

The power of the laser beam, L_p, is given by

$$L_p = \frac{E_s}{\Delta t}, \tag{13.3}$$

where E_s is the laser energy (in the unit of J) and Δt is the pulse duration of the laser. The power density of the laser beam, P_d (in the unit of W/mm^2), is given by

$$P_d = \frac{L_p}{A_s} = \frac{4L_p}{\pi (F_l \theta)^2}. \tag{13.4}$$

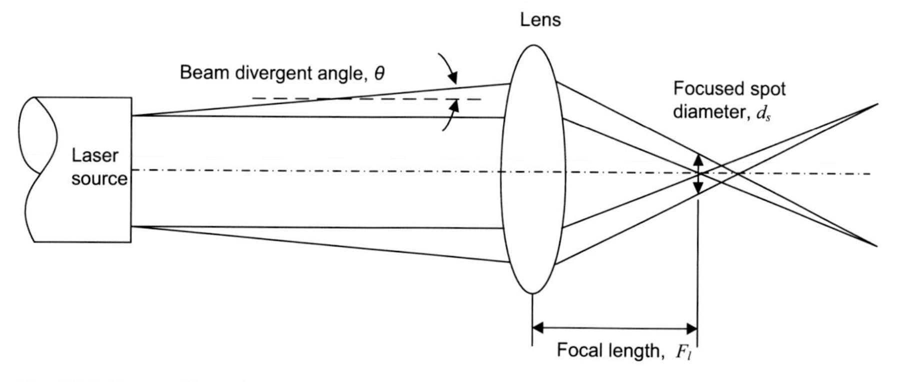

Fig. 13.2 Focus of laser beam

LBM occurs when the power density of the beam is greater than what is lost by thermal conduction, convection, and radiation.

The drilling feed rate f (in the unit of mm/s) can be described as follows:

$$f = \frac{C_l L_p}{E_v A_s} = \frac{C_1 P_d}{E_v}, \tag{13.5}$$

where the conversion efficiency C_l is a constant depending on the material and conversion efficiency and E_v is vaporization energy of the workpiece material (J/mm³). The MRR can be calculated as follows:

$$\text{MRR} = fA_s = \frac{C_l L_p}{E_v}. \tag{13.6}$$

Example
A pulsed laser beam machine is used to drill microholes. This machine generates laser pulses with 1 J/pulse energy and 0.001 s pulse duration. The beam divergence of the laser is 0.002 rad, and its radiation is focused with a 25 mm focal length lens. The vaporization energy of the material is 32 J/mm³. Suppose the conversion efficiency of input energy to thermal energy is 0.5 %, which corresponds to material removal. What is the diameter of the microhole? If the duty cycle of the laser pulse is 25 %, how fast is the machining rate f (mm/s)?

Solution
The size of the spot diameter d_s is determined by

$$d_s = F_1\theta = 25 \times 0.002 = 0.05 \text{ mm}.$$

The power of the laser beam, L_p, is:

$$L_p = \frac{E_s}{\Delta t} = \frac{1}{1 \times 10^{-3}} = 1 \text{ kW}.$$

The power density of the laser beam, P_d (W/mm²), is calculated by

$$P_d = \frac{L_p}{A_s} = \frac{4L_p}{\pi (F_1\theta)^2} = \frac{4 \times 1}{\pi (0.05)^2} = 509.3 \text{ kW/mm}^2.$$

The machining rate f (mm/s) can be obtained by

$$f = \frac{C_1 P_d}{E_v} = \frac{0.005 \times 509.3 \times 10^3}{32} = 79.58 \text{ mm/s}.$$

With 25 % duty cycle, the machining rate is:

$$f_{25\%} = 79.58 \times 0.25 = 19.9 \text{ mm/s}$$

Example

The laser absorption rate will decrease due to the plasma shielding effect. Given the same laser machine as in the previous example (1 J/pulse, 0.001 pulse duration, 25 % duty cycle, and lens with 25 mm focal length), an experiment is performed to study the plasma shielding effect. The experiment compares the time needed to drill holes 2.5 and 5 mm deep on a titanium workpiece with vaporization energy of 45 J/mm^3. The drilling time for 2.5 and 5 mm holes are 0.15 and 0.37 s, respectively. What is the energy conversion efficiency for each hole? (Assuming the hole has constant diameter.)

Solution

The power of the laser beam is $L_p = \frac{E_s}{\Delta t} = \frac{1}{1\times 10^{-3}} = 1 \text{ kW} = 1000 \text{ W}$.

For the 2.5 mm deep hole, the total input energy is

$$L_p \times \text{time} \times \text{duty cycle} = 1000 \times 0.15 \times 25\% = 37.5 \text{ J}.$$

The energy needed to vaporize titanium with the volume of the hole is:

$$E_v \times \frac{1}{4}\pi(F_1\theta)^2 \times \text{depth} = 45 \times \frac{1}{4}\pi(25 \times 0.002)^2 \times 2.5 = 0.22 \text{ J}.$$

Conversion efficiency *(η)* is:

$$\eta = \frac{0.22}{37.5} = 0.59\%.$$

For the 5 mm deep hole, the total input energy is:

$$L_p \times \text{time} \times \text{duty cycle} = 1000 \times 0.37 \times 25\% = 92.5 \text{ J}.$$

The energy needed to vaporize titanium with the volume of the hole is:

$$E_v \times \frac{1}{4}\pi(F_1\theta)^2 \times \text{depth} = 45 \times \frac{1}{4}\pi(25 \times 0.002)^2 \times 5 = 0.44 \text{ J}.$$

Conversion efficiency *(η)* is:

$$\eta = \frac{0.44}{92.5} = 0.48\%.$$

Table 13.4 Advantages and limitations of LBM

Advantages	Limitations
Capable of machining a wide variety of metallic and non-metallic materials	High equipment cost
	Occupational safety concern, particularly for eyes of operators
Low mechanical force and deformation of the workpiece	Limited dimensional and form accuracy and surface quality.
No diffraction and capable of simultaneously working at different workstations	Difficult to machine blind holes with precise depth
Generating micro features in difficult-to-machine and refractory materials	Tapered hole geometry
Capable of controlling the beam characteristics to adapt to a specific machining needs	Low machining efficiency (η commonly less than 1 %)
	Difficult to avoid the HAZ
No requirement for time-consuming vacuum as in EBM	Recast layer needed to be removed

HAZ heat-affected zone, *EBM* electron beam machining

In summary, LBM is a machining process that uses light as an energy source to remove materials. A wide variety of applications from welding to machining, from micro- to macro- to nanosize, and from energy to health care can be achieved by LBM. Table 13.4 gives a summary of advantages and limitations of LBM.

13.2 Electron Beam Machining

The basic principle of EBM is the melting and vaporizing of work material by the thermal energy generated by the electron beam. The electron beam in this process is generated within a well-controlled vacuum chamber (10^{-4}–10^{-1} Pa), similar to a cathode ray tube (CRT). Figure 13.3 shows a cathode, usually made of a tungsten filament, which is heated to 2500–3000 °C to emit the electrons that are accelerated by a high voltage (on the order of 50–200 kV) to 50–80 % of the speed of light and focused to a 0.25–1 mm diameter beam of an energy density over 1–2 MW/mm^2. Upon impact with the workpiece, the kinetic energy of electrons is transformed (at a high-energy-conversion efficiency close to 70 %) into thermal energy, sufficient to overcome the latent heat of the workpiece thereby locally melting and vaporizing the workpiece material. By virtue of the high energy involved, almost all engineering materials can be machined by this technique.

To control the HAZ, the temperature of the workpiece outside the cutting region is reduced by pulsing the electron beam. Pulse frequencies seldom exceed 10^4 Hz, and pulse duration is often of 0.05–100 ms. The resulting HAZ is narrow, therefore the machined cavity is typically small and deep with a depth-to-diameter ratio as high as 15. Enclosing the workpiece in high vacuum can achieve deep penetration,

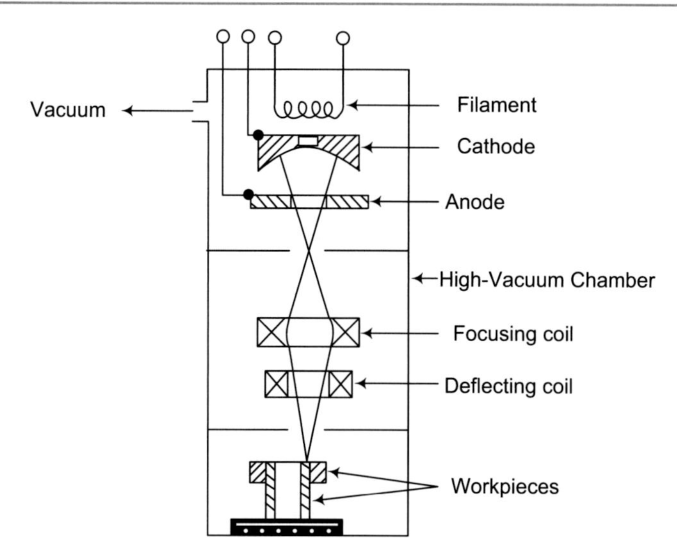

Fig. 13.3 High energy densities obtained in electron beam cutting

Table 13.5 History of electron beam machining (EBM)

Year	Event
1858	Electron was found from glow
1885	Electron was found from X-ray
1938	EBM in melting/vaporizing material
1950	Reached adequately high intensity 10^8 W/cm^{2}for industrial applications
1958	Steigerwald designed a prototype electron beam equipment that has been built by Messer-Griessheim in Germany for welding applications
1959	Used in micromachining
1967	Scanning EBM
1975	Electron-beam lithography for additive manufacturing

although the time it takes to pump down the workpiece chamber is a negative factor to productivity. The electron beam can be deflected with an electromagnetic coil. EBM can cut complex patterns in a highly automated manner (Table 13.5).

A machining system of EBM consists of the following four components:

1. Electron beam gun: This is a high-voltage source, for example, 120 kV, necessary to accelerate the electrons from the cathode (heated tungsten filament) toward the hollow anode, and the electrons continue their motion in vacuum toward the workpiece. The bias cup located between the cathode and the anode acts as a grid

that controls the beam current (1–80 mA) by controlling the number of electrons. Bias cup also acts as a switch for the electron beam to generate the pulse.

2. Vacuum chamber: The electron beam and workpiece are confined in the vacuum chamber to prevent the oxidation of filament and other elements, the loss of electron's kinetic energy due to collision with the massive molecules of air (O_2 and N_2), and contamination of metal vapor and debris. Loading and unloading the workpiece will take time and consume energy to pump the air and create a vacuum environment for EBM.
3. Magnetic lens: It focuses the electron beam on a spot of diameter ranging from 12 to 25 μm. The control of the beam direction is based on the Lorentz force law:

$$\boldsymbol{F} = q[\boldsymbol{E} + (v \times \boldsymbol{B})], \tag{13.7}$$

where $\boldsymbol{F}$ is the Lorentz force vector on the electron, $\boldsymbol{E}$ and $\boldsymbol{B}$ are the electrical and magnetic field, respectively, and $\boldsymbol{v}$ is the velocity vector of the electron. It is the same principle that is used in cathode ray tube (CRT) televisions.

4. Deflection coils: They are used to deflect the beam within a shallow angle to extend the machining range. The beam loses its speed and circularity. Through beam deflection, specific shapes and configurations can be produced.

The key operation parameters of EBM performance include the thermal properties of the workpiece (such as thermal conductivity, specific heat, and melting point), accelerating voltage (in kV), electron beam current (in mA), pulse energy, pulse duration, pulse frequency, spot diameter, and traverse speed of the workpiece. It has a similarity to LBM. The key difference is the need for a vacuum chamber to host the workpiece and electron beam source.

There are a variety of applications of EBM which include drilling, welding, and polishing. Advanced EBM system can perform scanning drilling, which synchronizes the motion of the electron beam and the workpiece. Workpiece manipulation occurs when the beam is deflected so that it moves in sequence with the part, thus allowing drilling while the part is moving. This is called an on-the-fly drilling operation. A more sophisticated multi-axis manipulator can be used for drilling holes in complicated shapes. The workpiece is held in a chuck while the motion of the axes is computer numerically controlled. The bearings must be carefully sealed to protect them from damage by metal vapor and drilling debris.

Advantages and limitations of EBM are summarized in Table 13.6.

A quantitative measure of the effect of electron emission is the emission current density (J_e) in the unit of A/mm^2 as influenced by the accelerating voltage (V_a) and the spacing between the anode and cathode (d_f) from :Childs–Langmuir relation:

$$J_e = k_e \frac{V_a^{3/2}}{d_f^{\,2}}, \tag{13.8}$$

Table 13.6 Advantages and limitations of electron beam machining (EBM)

Advantages	Limitations
Machining any material independent of its mechanical properties	High capital cost of equipment
Micromachining economically at higher speeds than that of EDM and ECM, e.g., applicable for drilling of fine holes at high rates (up to 4000 holes/s)	
Capability of maintain high accuracy and repeatability of ±0.1 mm for position and ±5% of the diameter of the drilled hole	Time loss for evacuating the machining vacuum chamber
No difficulty with acute angles	Presence of a thin recast layer and HAZ
Producing good surface finish	Necessity for auxiliary backing material
Providing a high degree of automation and productivity	

EDM electrical discharge machining, *ECM* electrochemical machining, *HAZ* heat-affected zone

where $k_e = 2 \sim 3 \times 10^6$ AV$^{-3/2}$. Typically, the emission current density can range from 0.01 to 1 A/mm^2, and is related to the current I_e through the expression of:

$$I_e = A_e J_e, \tag{13.9}$$

where A_e is the spot area of the electron beam.

The energy E_e of a single pulse of duration (t_p) is:

$$E_e = V_a I_e t_p, \tag{13.10}$$

and it determines the volume of material removal (m_e)

$$E_e = k m_e, \tag{13.11}$$

where k is a constant reflecting the energy-conversion efficiency and material properties, primarily the boiling point temperature and thermal conductivity. The value of k for aluminum is 360 J/mm^3 and for tungsten is 850 J/mm^3.

In electron beam drilling, the depth of eroded material per pulse h_e can be estimated:

$$h_e = \frac{4m_e}{\pi d_b^2} = \frac{4E_e}{k\pi d_b^2}, \tag{13.12}$$

where d_b is the diameter of the electron beam at contact with the workpiece.

The machining rates are usually evaluated in terms of the number of pulses required to evaporate a particular amount of material. Electron counters are used to register the number of pulses for easy adjustment of the machining time to produce a required depth of cut. The number of pulses (n_e) required for a cavity depth of h is:

$$n_{\mathrm{e}} = \frac{h}{h_{\mathrm{e}}} = \frac{hk\pi d_{\mathrm{b}}^2}{4E_{\mathrm{e}}}. \tag{13.13}$$

Example
An electron beam machine of 70 kV accelerating voltage and 0.05 A/mm^2 current density is used to drill a tungsten workpiece. Estimate the depth of the hole in 20 s. k_e is 850 J/mm^3 for tungsten.

Solution

$$h_e = \frac{m_e}{A_e} = \frac{E_e}{A}\left(\frac{1}{A_e}\right) = \frac{V_a I_e t_p}{kA_e} = \frac{V_a J_e A_e t_p}{kA_e} = \frac{V_a J_e t_p}{k} = \frac{(70\times10^3)(0.05)(20)}{850} = 82 \text{ mm}$$

Homework
A pulsed laser beam machine is used to drill microholes on zirconium (vaporization energy 42.5 J/mm^3). This machine generates laser pulses with 1.5 J/pulse energy and 0.001 s pulse duration. The beam divergence of the laser is 0.002 rad, and its radiation is focused with a 30 mm focal length lens. Suppose the conversion efficiency of input energy to thermal energy is 0.5 %, which corresponds to material removal. How many pulses are needed to drill a 1 mm deep hole?

14 Biomedical Machining

Machining is widely utilized in health care and is critical to the efficacy and outcome of the procedure. For example, needle insertion for vessel access, drug delivery, or biopsy is a tissue cutting and biomedical machining process. Dental drilling and root canal require extensive milling, grinding, and drilling processes. Sawing the plaster casts using the multi-points saw blade with oscillatory axial motion is a unique experience for many. The saw can cut through the hard plaster without damaging the soft skin underneath. Cutting occurs in clipping fingernails and toenails. Abrasive tools are often used for filing the wart and in skin care. Tooth brushing is a chemical mechanical polishing process. There are many biomedical machining procedures in surgery. One of the most commonly used surgical devices for cutting and coagulation of tissue is the monopolar electrosurgical device, which has the same basic principle and generator as in electrical discharge machining. In cataract surgery, a very sharp diamond scalpel is critical for ophthalmologists to cut the side of the cornea for lens replacement. Orthopedic surgeons use drills and various bone-cutting tools extensively for fixation of broken bones and for installing artificial implants for hip or knee joint replacement. Neurosurgeons perform very dedicated bone grinding and bipolar electrosurgical operations on the spine and brain for dissection of tumors.

Tissue is the workpiece in biomedical machining. The work material ranges widely; from soft tissue to hard bone and tooth. Biomedical machining is closely related to medical devices, which, similar to the machine tools, perform machining tasks and are critical to the efficacy of the procedure. Among many topics of biomedical machining, the oblique cutting of soft tissue, needle insertion, biomedical grinding, and electrosurgical processes are four topics addressed in this chapter.

S. Y. Liang, A. J. Shih, *Analysis of Machining and Machine Tools*,
DOI 10.1007/978-1-4899-7645-1_14

14.1 Oblique Cutting of Soft Tissue

Cutting of soft tissue, which includes skin, muscle, fat, cartilage, tendon, cornea, and others, is technically challenging. Lacking adequate support, the soft tissue deforms and limits the magnitude of cutting force for efficient cutting. Deformation is the most significant challenge in cutting soft tissue. Without adequate force, the soft tissue will continue to deform without cutting or separating from the workpiece. If the material is soft, ductile, and tough, it is difficult for the tool to effectively cut the workpiece. This phenomenon is evident at the dining table while using a knife to cut meat. If the meat is soft and easy to cut, the knife will cut straight into the meat, as illustrated in Fig. 14.1a and b.

Figure 14.1a shows the cross-sectional view of a double-beveled knife with a bevel angle ξ and a straight cutting edge. The knife cuts straight into the soft tissue in the orthogonal cutting configuration with the cutting edge (marked as vector $\boldsymbol{s}$) perpendicular to the cutting direction (marked as vector $\boldsymbol{v}$), as shown in Fig. 14.1b. This orthogonal cutting is also illustrated in Fig. 14.2a with the cutting edge $\boldsymbol{s}$ on the tool's cutting face A_c (with normal vector $\boldsymbol{n}$) and the cutting direction $\boldsymbol{v}$ perpendicular to $\boldsymbol{s}$.

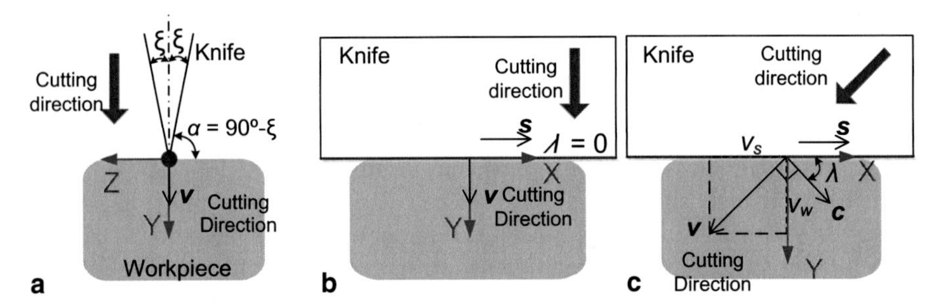

Fig. 14.1 The straight cutting edge of a knife tool. **a** *Side view* and the bevel angle (ξ) and rake angle (α), **b** *front view* of the orthogonal cutting with the cutting edge $\boldsymbol{s}$ perpendicular to the cutting direction $\boldsymbol{v}$, and **c** *front view* of the oblique cutting with the inclination angle (λ)

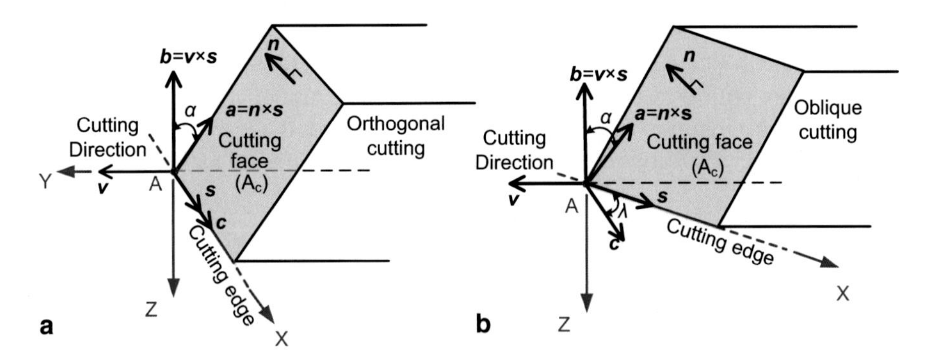

Fig. 14.2 **a** Orthogonal cutting and the rake angle and **b** oblique cutting and the inclination angle

If the meat is tough and difficult to cut, we intuitively oscillate the knife back and forth to cut the meat. This is illustrated in Fig. 14.1c. The knife motion creates the oblique cutting configuration and introduces the inclination angle λ, as shown in Figs. 14.1b and c. Three vectors, $\boldsymbol{a}=\boldsymbol{n}\times\boldsymbol{s}$, $\boldsymbol{b}=\boldsymbol{v}\times\boldsymbol{s}$, and $\boldsymbol{c}=\boldsymbol{b}\times\boldsymbol{v}$ are introduced to define the rake and inclination angles in Figs. 14.1 and 14.2.

- Rake angle α is the angle between vectors $\boldsymbol{a}$ and $\boldsymbol{b}$.
- Inclination angle λ is the angle between vectors $\boldsymbol{s}$ and $\boldsymbol{c}$.

In orthogonal cutting configuration (Figs. 14.1b and 14.2a), $\lambda=0$.

In Fig. 14.1a, an XYZ coordinate is defined on a double-bevel knife edge with the X axis along the cutting edge and Z axis perpendicular to the symmetry plane of the knife. For straight cutting into the workpiece (Fig. 14.1b), $\boldsymbol{c}$ is perpendicular to $\boldsymbol{v}$ and in the same direction as $\boldsymbol{s}$. The inclination angle $\lambda=0°$ and rake angle $\alpha=90°-\xi$.

$$\begin{aligned}
\boldsymbol{s} &= [1,0,0] \\
\boldsymbol{v} &= [0,1,0] \\
\boldsymbol{n} &= [0,\cos\alpha,-\sin\alpha]=[0,\sin\xi,-\cos\xi] \\
\boldsymbol{a} &= [0,-\sin\alpha,-\cos\alpha]=[0,-\cos\xi,-\sin\xi] \\
\boldsymbol{b} &= [0,0,-1] \\
\boldsymbol{c} &= [1,0,0].
\end{aligned}$$

For oblique cutting using the knife (Fig. 14.1c), there are two velocity components: along the cutting edge v_s and into the workpiece v_w. In this case, the rake angle $\alpha=90°-\xi$. The inclination angle is

$$\lambda=\arctan\left(v_s/v_w\right) \tag{14.1}$$

and

$$\begin{aligned}
\boldsymbol{s} &= [1,0,0] \\
\boldsymbol{v} &= [-v_s,v_w,0] \\
\boldsymbol{n} &= [0,\sin\xi,-\cos\xi] \\
\boldsymbol{a} &= [0,-\cos\xi,-\sin\xi] \\
\boldsymbol{b} &= [0,0,-v_w] \\
\boldsymbol{c} &= [v_w^2,v_sv_w,0].
\end{aligned}$$

Example

While cutting the meat, the speed of the knife with 5° bevel angle has $v_s=2$ mm/s and $v_w=2$ mm/s. What is the inclination angle in this oblique cutting setup? It is common that when the meat is tough to cut, v_s is increased intuitively by the knife user. Assuming v_s is increased to 5 mm/s and v_w remains at 2 mm/s, what is the inclination angle for cutting? Draw an oblique cutting configuration based on these two inclination angles. Find the rake angle for these two cutting configurations.

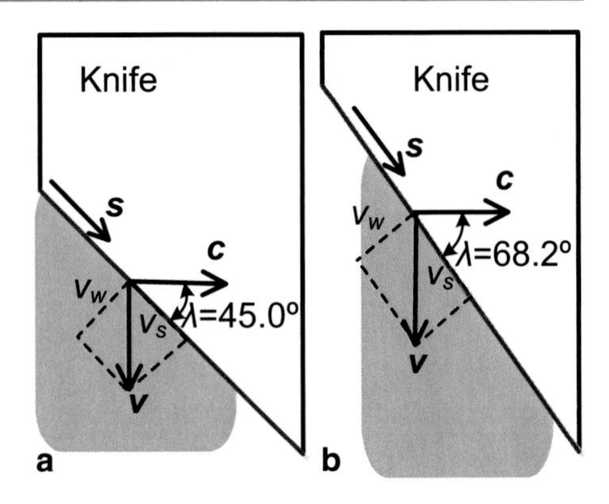

Fig. 14.3 Two oblique cutting setups. **a** $v_s = v_w = 2$ mm/s and **b** $v_s = 5$ mm/s and $v_w = 2$ mm/s

Solution
For $v_s = v_w = 2$ mm/s and $\xi = 5°$, $\boldsymbol{s} = [1, 0, 0]$, $\boldsymbol{v} = [-2, 2, 0]$, $\boldsymbol{n} = [0, 0.087, -0.996]$, $\boldsymbol{a} = [0, -0.996, -0.087]$, $\boldsymbol{b} = [0, 0, -2]$, and $\boldsymbol{c} = [4, 4, 0]$. The inclination angle $\lambda = \tan^{-1}(1) = 45°$. The rake angle $\alpha = 90° - \xi = 85°$.

For $v_s = 5$ mm/s, $\boldsymbol{s} = [1, 0, 0]$, $\boldsymbol{v} = [-5, 2, 0]$, $\boldsymbol{n} = [0, 0.087, -0.996]$, $\boldsymbol{a} = [0, -0.996, -0.087]$, $\boldsymbol{b} = [0, 0, -2]$, and $\boldsymbol{c} = [4, 10, 0]$. The inclination angle $\lambda = \tan^{-1}(2.5) = 68.2°$. The rake angle α remains the same as 85° (Fig. 14.3).

Soft tissue cutting studies have shown that the inclination angle is important in the initial cutting force. Once the started of soft tissue cutting, the effect of rake angle becomes important. The sliding motion along the cutting edge can increase the inclination angle and reduce the initial cutting force. It is common that the knife cutting edge is not straight. This will change the inclination angle at the cutting point. Examples of the non-straight cutting edge include the serrated bread knife and sawtooth cutting edges.

Example
Roller cutter is commonly used for slicing pizza. Assume using a 4 in. diameter circular cutter to cut a 1 in. thick pizza, as shown in Fig. 14.4. Find the inclination angle at three points, marked as A_1, A_2, and A_3, during the straight down cutting into pizza. Point A_1 is at the first contact point with pizza in straight down cutting (Fig. 14.4a), A_2 is 30° from A_1, and A_3 is on the top edge of the pizza after cutting down 1 in. Assume the circular cutter is in pure rolling without sliding across the pizza (Fig. 14.4b), find the inclination angle at points A_2 and A_3 during the rolling cutting of pizza.

Solution
As shown in Fig. 14.5a, during the straight cutting, $\lambda = 0°$ at A_1, $\lambda = 30°$ at A_2, and $\lambda = 60°$ at A_3. During the rolling cutting, Fig. 14.5b, $\lambda = 15°$ at A_2 and $\lambda = 30°$ at A_3.

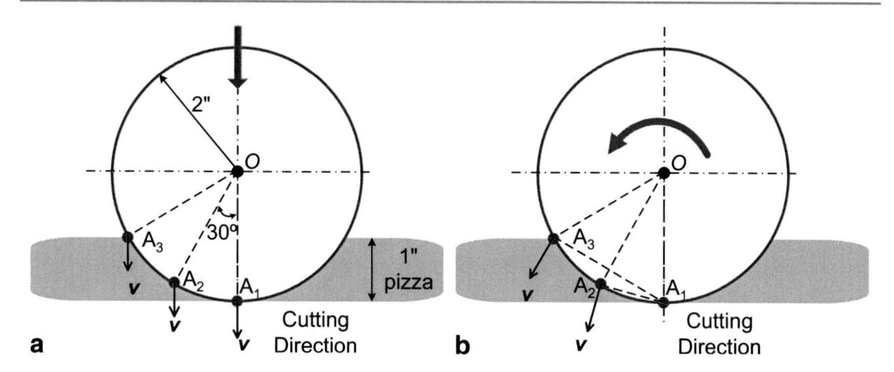

Fig. 14.4 Schematic view of two cutting setups for roller cutter to cut pizza. **a** Straight down cutting and **b** pure rolling cutting

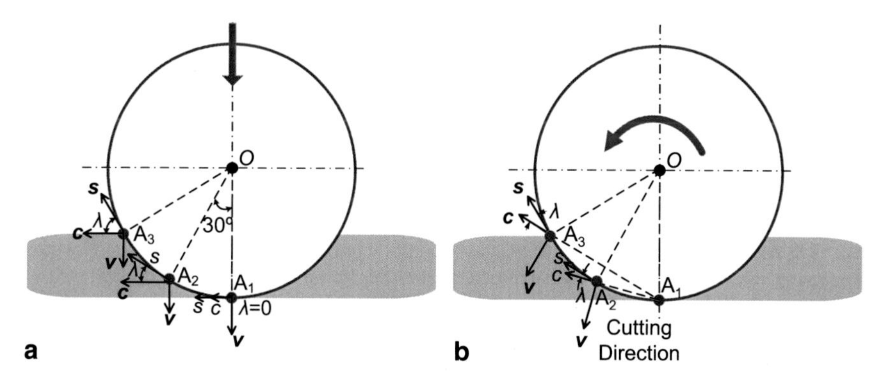

Fig. 14.5 Inclination angles of two cutting setups for roller cutter to cut pizza. **a** Straight down cutting and **b** pure rolling cutting

14.2 Needle Cutting Edges

A needle is the most common feature in medical devices. Needle insertion is a tissue-cutting process. Needle tip geometry determines the cutting edge and its rake and inclination angles and the insertion force. Guidance and biopsy are two functions of the needle that are key to the design of needle cutting edge to meet the performance requirements. Analyzing the needle cutting edge and the corresponding rake and inclination angles is fundamental in needle design.

The simplest type of needle is the single-plane bias bevel needle, as shown in Fig. 14.6a. This needle is fabricated by grinding a single plane with bevel angle ξ on a needle tube, typically made of stainless steel, with the inside radius r_i and outside radius r_o. Two cutting edges of the single-plane bevel needle are both semielliptical curves as marked in Fig. 14.6a. In the XYZ coordinate system, the Z axis coincides with the central axis of the needle tube and the X axis passes through the lowest point of the needle tip profile.

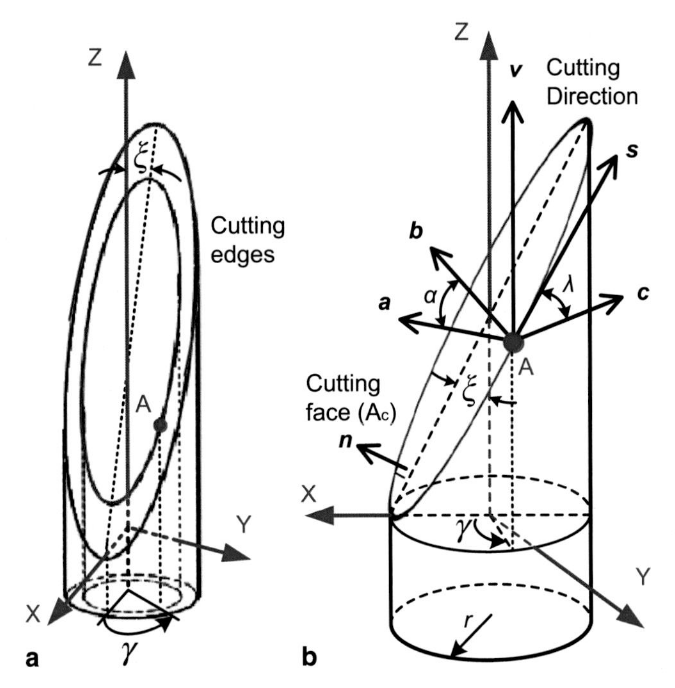

Fig. 14.6 **a** Single-plane bias bevel needle and definition of XYZ coordinate system and **b** vectors ***v***, ***n***, ***s***, ***a***, ***b***, and ***c*** and the inclination and rake angles defined on an elliptical cutting edge

Commercial needles are manufactured by grinding of planes on the needle tip. Therefore, the needle cutting edges commonly consist of segments of elliptical curves. To study the cutting edges, the elliptical cutting edges in Fig. 14.6b and the geometric relationship of vectors ***v***, ***n***, ***s***, ***a***, ***b***, and ***c*** are studied. The XYZ coordinate of the point A on the cutting edge in a semielliptical curve with radius r of the base cylinder is defined by the angle γ measured from the X axis.

$$\begin{aligned} A_x &= r\cos\gamma \\ A_y &= r\sin\gamma \\ A_z &= r(1-\cos\gamma)\cot\xi. \end{aligned} \tag{14.2}$$

The vector ***s*** at point A in Fig. 14.6a is the derivative of $[A_x, A_y, A_z]$ with respect to γ.

$$\boldsymbol{s} = [-\sin\gamma,\ \cos\gamma,\ \cot\xi\sin\gamma]. \tag{14.3}$$

The vector ***n*** normal to the cutting plane A_c is $[\cos\xi, 0, \sin\xi]$. The cutting direction $\boldsymbol{v}=[0, 0, 1]$. Based on ***n***, ***v***, and ***s***, vectors ***a***, ***b***, and ***c*** can be derived.

$$\boldsymbol{a} = \boldsymbol{n} \times \boldsymbol{s} = [-\cos\gamma \sin\xi, -\sin\gamma \sin\xi - \cos\xi \cot\xi \sin\gamma, \cos\gamma \cos\xi]$$
$$\boldsymbol{b} = \boldsymbol{v} \times \boldsymbol{s} = [-\cos\gamma, -\sin\gamma, 0]$$
$$\boldsymbol{c} = \boldsymbol{b} \times \boldsymbol{v} = [-\sin\gamma, \cos\gamma, 0].$$

The rake angle is:

$$\alpha = \cos^{-1}\frac{\boldsymbol{a}\cdot\boldsymbol{b}}{|\boldsymbol{a}||\boldsymbol{b}|} = \cos^{-1}\sqrt{\cos^2\gamma \sin^2\xi + \sin^2\gamma}. \tag{14.4}$$

The inclination angle λ is

$$\lambda = \cos^{-1}\frac{\boldsymbol{s}\cdot\boldsymbol{c}}{|\boldsymbol{s}||\boldsymbol{c}|} = \cos^{-1}\frac{1}{\sqrt{1+\cot^2\xi \sin^2\gamma}}. \tag{14.5}$$

Example Calculate the inclination and rake angles at three points A_1, A_2, and A_3 with $\gamma = 180$, 225, and 270°, respectively, of the single-plane bias bevel needle with 30° bevel angle. Plot the inclination and rake angles for this needle with bevel angle of 10, 20, and 30° and outside radius of 1 (unit length).

Solution

In Fig. 14.7, the needle insertion direction $\boldsymbol{v} = [0, 0, 1]$ and the cutting surface $\boldsymbol{n} = [0.866, 0, 0.5]$.

At point A_1, $\gamma = 180°$ (tip of the needle), the XYZ coordinates are $(-1, 0, 3.464)$, and $\boldsymbol{s} = [0, -1, 0]$.

$$\boldsymbol{a} = [0.5, 0, -0.866]$$
$$\boldsymbol{b} = [1, 0, 0]$$
$$\boldsymbol{c} = [0, -1, 0],$$

$$\alpha = \cos^{-1}\frac{\boldsymbol{a}\cdot\boldsymbol{b}}{|\boldsymbol{a}||\boldsymbol{b}|} = \cos^{-1}\sqrt{\cos^2\gamma \sin^2\xi + \sin^2\gamma} = 90^\circ - \xi = 60^\circ,$$

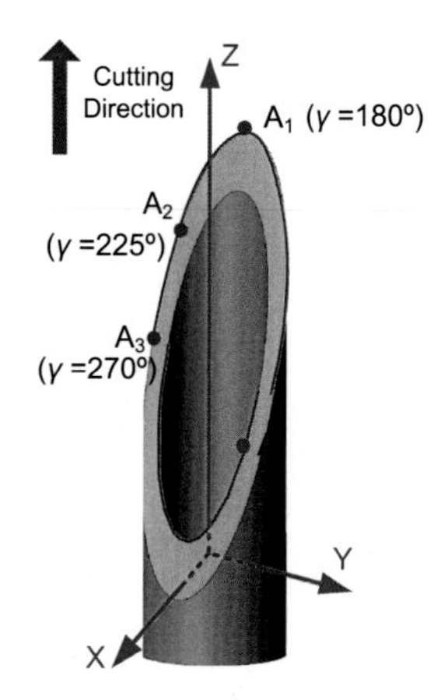

Fig. 14.7 Single-plane bias bevel needle and definition of XYZ coordinate system

$$\lambda = \cos^{-1}\frac{\boldsymbol{s}\cdot\boldsymbol{c}}{|\boldsymbol{s}||\boldsymbol{c}|} = \cos^{-1}\frac{1}{\sqrt{1+\cot^2\xi\sin^2\gamma}} = \cos^{-1}(1) = 0^{\circ}.$$

At point A_2, γ=225°, the XYZ coordinates are: (−0.707, −0.707, 2.957) and $\boldsymbol{s}$=[0.707, −0.707, −1.225].

$$\begin{aligned}\boldsymbol{a} &= [0.354, 1.414, -0.612]\\ \boldsymbol{b} &= [0.707, 0.707, 0]\\ \boldsymbol{c} &= [0.707, -0.707, 0],\end{aligned}$$

$$\alpha = \cos^{-1}\frac{\boldsymbol{a}\cdot\boldsymbol{b}}{|\boldsymbol{a}||\boldsymbol{b}|} = \cos^{-1}\sqrt{\cos^2\gamma\,\sin^2\xi + \sin^2\gamma} = \cos^{-1}\sqrt{0.625} = 37.8^{\circ},$$

$$\lambda = \cos^{-1}\frac{\boldsymbol{s}\cdot\boldsymbol{c}}{|\boldsymbol{s}||\boldsymbol{c}|} = \cos^{-1}\frac{1}{\sqrt{1+\cot^2\xi\sin^2\gamma}} = \cos^{-1}\left(\frac{1}{\sqrt{2.5}}\right) = 50.8^{\circ}.$$

At point A_3, γ=270°, the XYZ coordinates are: (0, −1, 1.732) and $\boldsymbol{s}$=[1, 0, −1.732].

$$\begin{aligned}\boldsymbol{a} &= [0, 2, 0]\\ \boldsymbol{b} &= [0, 1, 0]\\ \boldsymbol{c} &= [1, 0, 0],\end{aligned}$$

$$\alpha = \cos^{-1}\frac{\boldsymbol{a}\cdot\boldsymbol{b}}{|\boldsymbol{a}||\boldsymbol{b}|} = \cos^{-1}\sqrt{\cos^2\gamma\,\sin^2\xi + \sin^2\gamma} = \cos^{-1}(1) = 0^{\circ},$$

$$\lambda = \cos^{-1}\frac{\boldsymbol{s}\cdot\boldsymbol{c}}{|\boldsymbol{s}||\boldsymbol{c}|} = \cos^{-1}\frac{1}{\sqrt{1+\cot^2\xi\sin^2\gamma}} = \cos^{-1}(0.5) = 60^{\circ}.$$

In summary, for the point A

$$\begin{aligned}\boldsymbol{s} &= [-\sin\gamma,\ \cos\gamma,\ \cot\xi\sin\gamma]\\ \boldsymbol{v} &= [0, 0, 1]\\ \boldsymbol{n} &= [\cos\xi, 0, \sin\xi]\\ \boldsymbol{a} &= [-\cos\gamma\,\sin\xi, -\sin\gamma\,\sin\xi - \cos\xi\,\cot\xi\,\sin\gamma,\ \cos\gamma\,\cos\xi]\\ \boldsymbol{b} &= [-\cos\gamma, -\sin\gamma, 0]\\ \boldsymbol{c} &= [-\sin\gamma, \cos\gamma, 0].\end{aligned}$$

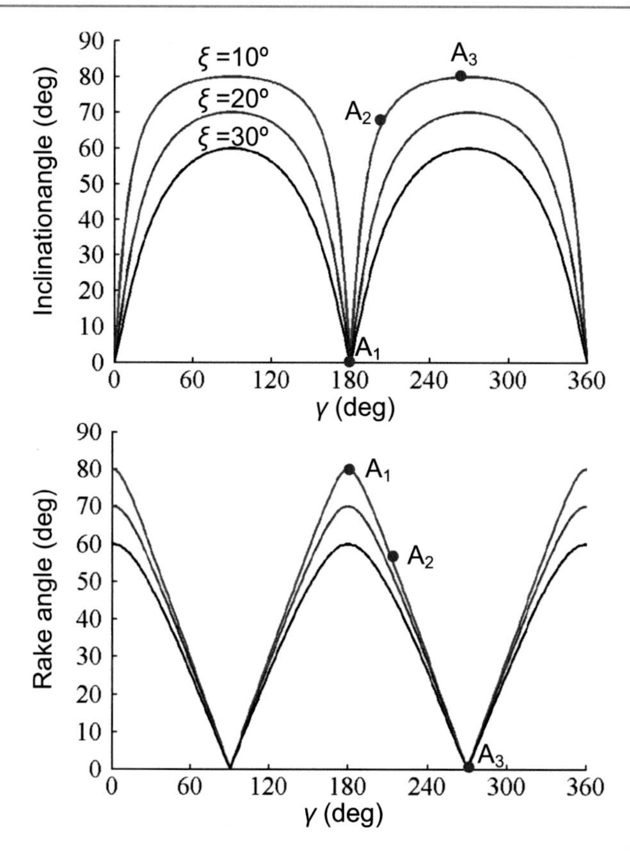

Fig. 14.8 Inclination and rake angles for single-plane bias bevel needle with bevel angle of 10, 20, and 30°

The inclination and rake angles for single-plane bias bevel needle with bevel angles of 10°, 20°, and 30° are plotted in Fig. 14.8.

Commercial needles look very different from the single-plane bias bevel needle shown in Fig. 14.6. By grinding two additional planes on the needle tip, the inclination angle can be increased (particularly at the needle tip) to reduce the initial needle insertion force. Point A_1 in the previous example has $\lambda = 0°$, which is the worst orthogonal cutting configuration for soft tissue cutting. Figure 14.9 shows two common commercial needles, namely the needle with lancet point (or commonly known as hypodermic needle or Chiba needle) and Franseen needle with three symmetry planes and their cutting edges and inclination angles. Figure 14.10 shows the rake and inclination angles of these three needles.

Fig. 14.9 **a** Single-plane bias bevel needle, and two common commercial needles: **b** the needle with lancet point and **c** Franseen (three-plane) needle

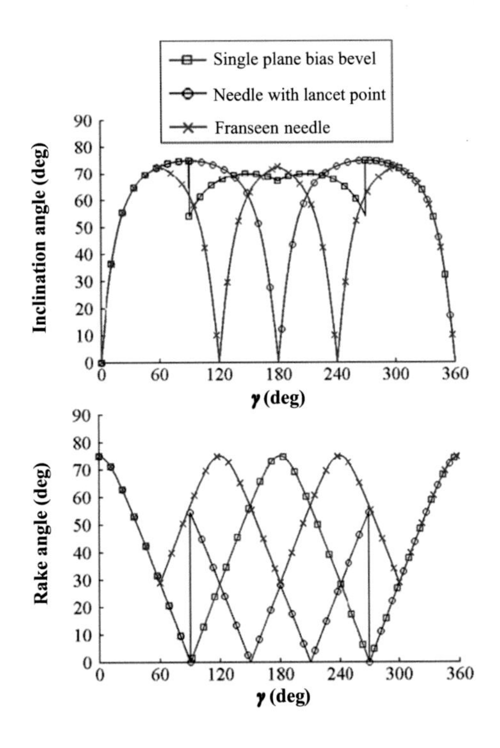

Fig. 14.10 Inclination and rake angles for three needles in Fig. 14.9

14.3 Force for Blade and Needle Insertion

As discussed in Section 14.1, due to the lack of structural support of soft tissue, large deformation is often associated with the cutting of soft tissue. During blade cutting or needle insertion of the soft tissue, the blade or needle cutting edge continues to deform the soft tissue workpiece with no penetration (or cutting). Figure 14.11 shows an experimentally measured insertion force versus insertion depth curve of an 11-gauge needle with ex-vivo porcine liver. The force gradually increases and

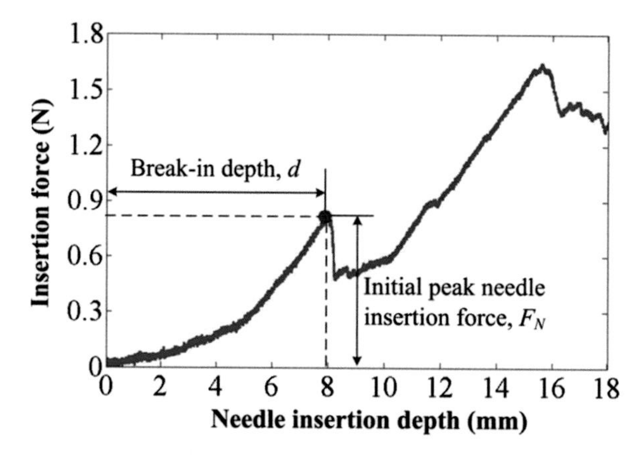

Fig. 14.11 A typical needle insertion force versus depth

the soft tissue deforms until the peak force has reached the break-in point, which has the break-in depth (d) and initial peak needle insertion force (F_N). The soft tissue has a large deformation (8 mm) and reaches the peak force of 0.84 N at this point before the start of tissue cutting. After the needle tip breaks into the soft tissue, the needle force drops suddenly and both the needle force and tissue deformation continue to increase until the next peak or the second break-in point.

The force versus depth graph for needle insertion into soft tissue shown in Fig. 14.12 is similar for the blade. Experiments using 16 blades with different rake and inclination angles for cutting of ex-vivo liver have been conducted to find the specific force model $f(\lambda,\alpha)$. The specific force is the blade cutting force divided by the projected length perpendicular to the cutting direction. Figure 14.12 shows the measured average $f(\lambda,a)$. The surface depicts $f(\lambda,a)$ (with unit of N/mm) that corresponds to a third-order multivariable (λ and a), best fit polynomial of the data, can be represented as:

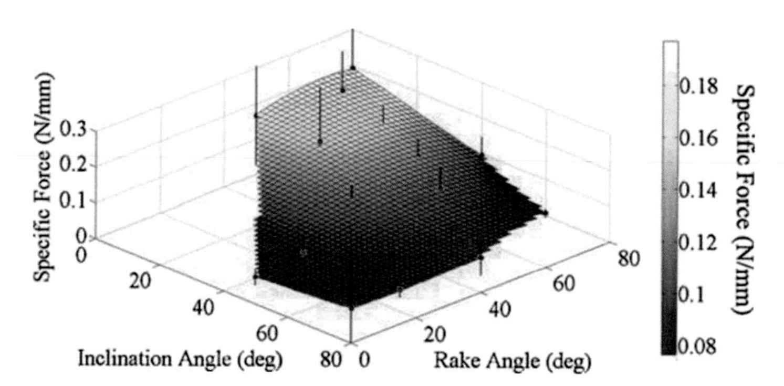

Fig. 14.12 The specific blade insertion force $f(\lambda,\alpha)$

$$f = -0.042 + 0.296\lambda + 0.298\alpha - 0.255\lambda^2 - 0.408\lambda \alpha - 0.011\alpha^2 + 0.083\lambda^3 + 0.118\lambda^2\alpha + 0.080\lambda\alpha^2 - 0.059\alpha^3 \quad (14.6)$$

The model fits well with experimental data ($R^2 = 0.97$). This $f(\lambda,\alpha)$ function can be used to analyze the force for cutting soft tissue during needle insertion as well as for optimal design of the needle shape with the lowest insertion force.

14.4 Needle Tip Grinding

Grinding is the process used to create the geometry of the needle tip. The grinding setup parameters define the needle tip geometry and will be used in the next section for the optimal design of the needle tip.

For example, the needle with lancet point is fabricated using the four-step grinding procedure shown in Fig. 14.13. In Step 1, the needle tube is tilted by the bevel angle ξ to grind the single-plane bias bevel needle tip at the end of the tube. In Step

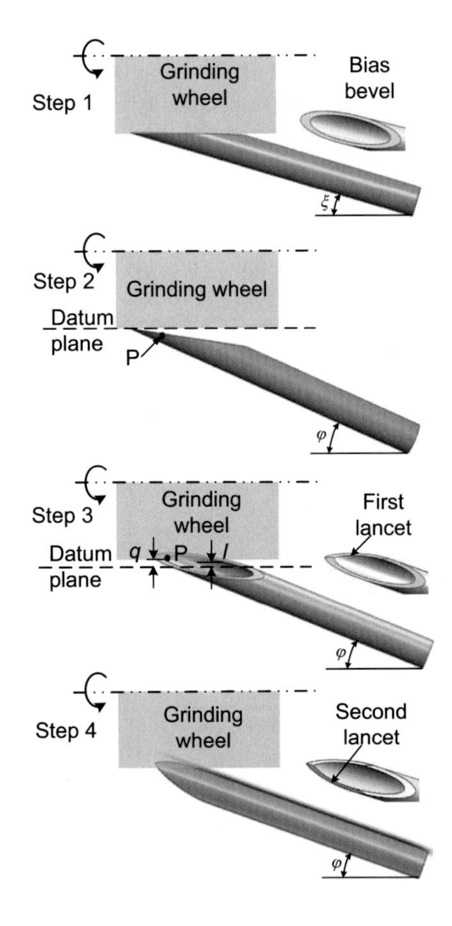

Fig. 14.13 Four steps in grinding the needle with lancet tip

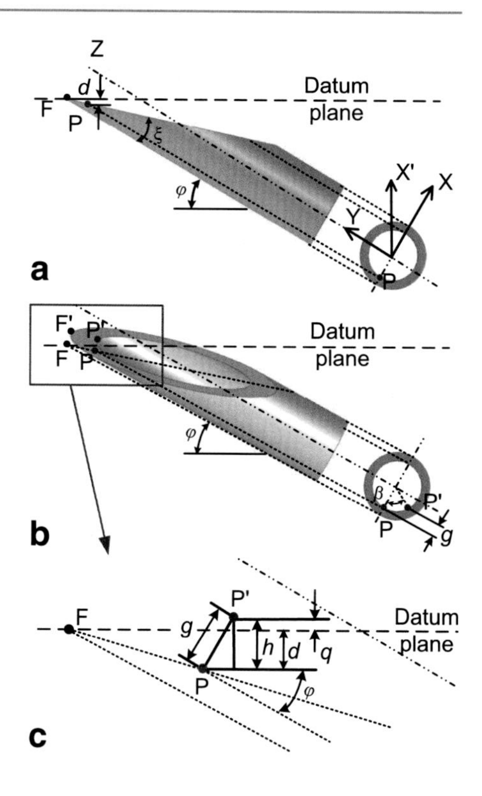

Fig. 14.14 Geometrical relationship to calculate q

2, the tilt angle is increased to φ, which is defined as the secondary bevel angle. The grinding wheel touches on the needle tip and defines the datum plane. In Step 3, the tube is rotated (or spun) along the centerline axis by an angle, denoted as β, and the grinding wheel is moved up by a distance, q, from the datum plane to grind the first lancet. In Step 4, the needle tube is rotated 2β in the direction opposite to the rotation in Step 3 and the second lancet is ground at the same height as in Step 3. In summary, three parameters for grinding a lancet needle tip are ξ, φ, and β. These three parameters define the shape of the lancet needle tip and will be used in the next section for the optimal design of needle.

The parameter q is calculated based on the geometrical relationship shown in Fig. 14.14, which illustrates the procedure to calculate q (Step 3 in Fig. 14.12) for grinding the lancet needle. In Fig. 14.14a, when the needle is oriented as in Step 2 (Fig. 14.13), the distance d from the datum plane (the needle tip touching the grinding wheel) to the point P in the X′ direction (perpendicular to the datum plane) is:

$$d = \frac{\sin(\varphi - \xi)(r_o - r_i)}{\sin\xi}, \tag{14.7}$$

where r_o and r_i are the outside and inside radius of the needle tube, respectively.

When the needle tube rotates by an angle of rotation, β, as shown in Fig. 14.14b, the point P moves to P′, with the distance $g = r_i(1-\cos\beta)$, as illustrated in Fig. 14.14b. The value of h, which is the distance from P to P′ along the X′ axis, is $g\cos\varphi$ and thus

$$h = r_i(1-\cos\beta)\cos\varphi. \tag{14.8}$$

The distance q between the datum plane to the new location of P is equal to $h-d$:

$$q = r_i(1-\cos\beta)\cos\varphi - \frac{\sin(\varphi-\xi)(r_o - r_i)}{\sin\xi} \tag{14.9}$$

Commercially, the outside diameter (OD) of the needle tube is defined by the gauge, ranging from gauge 6 (0.203 in. nominal diameter) to gauge 36 (0.00425 in. nominal diameter). Gauge 18 (0.05 in. nominal diameter) is one of the most common needles in clinic. The needle's inside diameter is determined by the wall thickness. Four needle tube wall thickness specifications, the regular wall (RW), thin wall (TW), extra thin wall (ETW), and ultra thin wall (UTW), are available commercially. For example, the wall thickness for 18-gauge RW, TW, ETW, and UTW needles are 0.0085, 0.006, 0.003, and 0.002 in., respectively. Data sheet is available for commercial needle tubes of various gauge and wall thicknesses.

Example

For an 11-gauge RW needle with lancet point and $\xi = 12°$, $\varphi = 18°$, and $\beta = 60°$, what is the q for setting up the grinding of the extra thin wall tube?

Solution

For the 11-gauge RW needle, the OD is 0.12 in. and wall thickness is 0.013 in. The $r_o = 0.06$ in. and $r_i = 0.047$ in. Based on Eq. (14.9), $q = 0.0158$ in. This is the needle to be used in the following section.

14.5 Optimal Needle with Lancet Point

Needle A, as shown in Fig. 14.15a, is a needle with lancet point widely used in health care. The bevel length, which is defined as the length of the bevel in the needle tip along the needle axis, of this 11-gauge needle tip with $\xi = 12°$, $\varphi = 18°$, and $\beta = 60°$ is 13.3 mm. Based on the specific cutting force function f in Fig. 14.12 and assuming the full contact area between the needle and the soft tissue (i.e., $\theta = 360°$ in Fig. 14.15), the mathematical model-predicted peak needle insertion force is 1.06 N.

Using the genetic algorithm for optimization and assuming $\theta = 360°$, the lancet needle with the lowest insertion force and the same bevel length has $\xi = 12.4°$, $\varphi = 12.4°$, and $\beta = 15.0°$, shown as the Needle B in Fig. 14.15a. The model-predicted needle insertion force is 0.94 N, 11 % lower than that of Needle A. By keeping the

Needles		Angles	Bevel length (mm)	Model-predicted needle insertion force (N)
Needle A	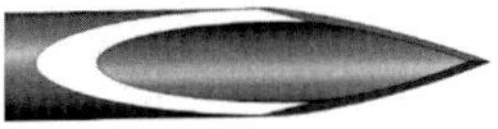	$\xi = 12.0°$ $\varphi = 18.0°$ $\beta = 60.0°$	13.30	1.06
Needle B		$\xi = 12.4°$ $\varphi = 12.4°$ $\beta = 15.0°$	13.30	0.94
Needle C		$\xi = 23.0°$ $\varphi = 23.0°$ $\beta = 10.0°$	7.17	1.06

a *Assuming $\theta = 360°$.

Needles		Angles	Bevel length (mm)	Contact angle, θ	Initial peak insertion force, F_N		
					Measured (N)	Predicted (N)	Discrepancy
Needle A	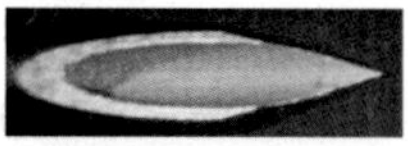	$\xi = 12°$ $\varphi = 18°$ $\beta = 60°$	13.30	287.9°	0.94	0.89	5.6%
Needle B	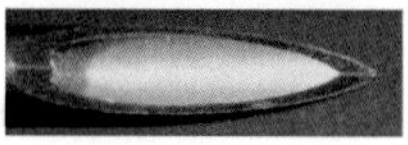	$\xi = 12°$ $\varphi = 12°$ $\beta = 15°$	14.29	216.2°	0.81	0.78	3.7%
Needle C	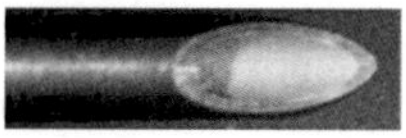	$\xi = 23°$ $\varphi = 23°$ $\beta = 10°$	7.17	360.0°	1.02	1.06	3.9%

b

Fig. 14.15 Optimal designed and manufactured needles with lancet point. Needle A (baseline needle), Needle B (minimal insertion force with same bevel length), and Needle C (minimal bevel length with same insertion force). **a** Optimized needle geometry and **b** manufactured needles

same force and minimizing the bevel length, the optimal needle, Needle C, has the bevel length of 7.17 mm bevel length, a 46 % reduction from that of Needle A. Shorter bevel length is beneficial for physicians and nurses to access the blood vessel.

For grinding the needles for experiment, since the index table has 1° resolution, some setup angles are adjusted or transacted to the nearest degree. The contact angle θ, as defined in Fig. 14.16, is not 360° in needle insertion experiment to the ex-vivo liver. The measured θ of three needles is shown in Fig. 14.15b. Based on the contact angle, the model-predicted and experimentally measured needle insertion forces are listed in Fig. 14.15b. Experimental measurements show that the initial peak insertion force for Needle B is reduced to 0.81 N, which is a 14 % reduction from the 0.94 N insertion force of Needle A. Needle C, which has the same 46 % reduction in bevel length, has 8.5 % higher insertion force (1.02 N) than that of Needle A. In summary, this example demonstrated that needle design and manufacturing are closely linked together. Three grinding setup parameters are used for optimal needle design. The optimal needle has either lower insertion force or shorter bevel length, or could be the combination of both.

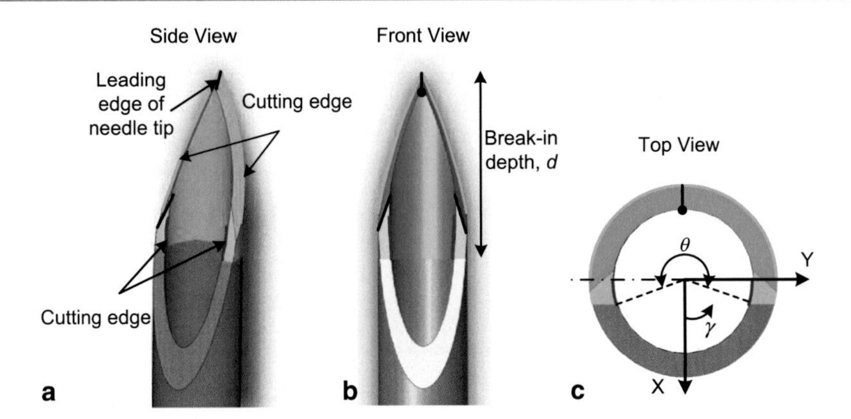

Fig. 14.16 The tissue contact region and angle θ. **a** *Side view*, **b** *front view*, and **c** *top view* of the lancet needle tip contact area with the soft tissue at the initial break-in point during insertion

14.6 Biomedical Grinding: Dental, Skull Base Neurosurgery, and Atherectomy

Abrasive processes are common in grinding teeth in dental procedures, bone in the skull base brain surgery, and plaque inside blocked coronary and peripheral artery.

Dental grinding typically uses metal-bond diamond or tungsten carbide cutting tools (also called bur) with 1.60 or 2.35 mm diameter shank. The dental grinding utilizes the high-speed (typically 300,000–400,000 rpm) wheel and compressed air-driven turbine spindle. This is the most common biomedical grinding process.

Bone grinding is also a critical surgical procedure for orthopedic surgery and neurosurgery. For example, a neurosurgical procedure for brain tumor in the base of the skull is called endoscopic endonasal approach to the skull base. This minimally invasive surgical procedure uses the nostril as a natural corridor to reach the interior of skull without disfiguring incisions. Figure 14.17 shows a commercial neurosurgical grinding device and two grinding wheels. Neurosurgeons use the miniature, 3–4 mm diameter, spherical diamond grinding wheel to remove the bone and gain the exposure for tumor removal. During bone grinding, neurosurgeons also need to identify and protect the important cranial nerves. The wheel rotational speed typically is over 50,000 rpm in such endoscopic surgical operations.

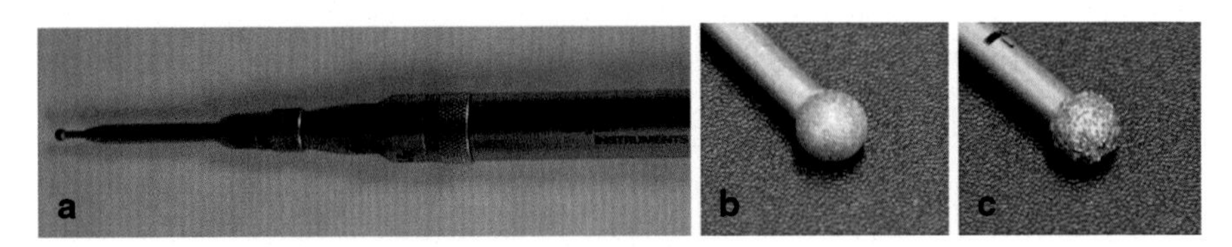

Fig. 14.17 **a** Neurosurgical grinding device and diamond wheel with **b** 25/30 ANSI mesh (711 FEPA), and **c** 200/230 ANSI mesh (76 FEPA) diamond grit

Heat is generated during bone grinding and propagates through the bone to the adjacent nerves and blood vessels. The rising temperature of the bone, nerve, and artery from bone grinding is a recognized phenomenon by surgeons and can cause three types of thermal-related injury. First, bone necrosis typically starts when the temperature is above a critical value of 50 °C. Second, nerve is vulnerable to elevated temperature. Depending on the type of nerve, the critical temperature for starting the thermal injury could start at 43 °C. Third, temperature rise can coagulate the blood and cause the blood to clot. For example, in endonasal approach to the skull base, coagulation of blood in the carotid artery due to the temperature rise caused by bone grinding can cause stroke. Exposing and resecting tumor lesions require extensive grinding of the bone surrounding the optic and cranial nerves to identify and preserve these critical nerves. Heat generated during skull base bone grinding may damage the nerve and cause blindness and the loss of facial muscle control. Heat is thus of particular concern for neurosurgeons.

Minimizing the temperature rise caused by bone grinding is the key to preventing the thermal injury in skull base bone grinding. For precision motion control and minimized trauma in bone grinding, some neurosurgeons prefer to use the grinding wheel with small grit size to feel a certain level of resistive force in bone grinding. Such higher grinding force converts to more heat generation during the neurosurgical bone grinding in the skull base. Currently, the primary mode of cooling is the saline irrigation during surgical bone grinding. The confined surgical space in skull base bone grinding and the high-speed wheel both conspire to limit the effectiveness of saline for cooling in bone grinding and open a lot of opportunities for advanced surgical grinding devices.

Grinding is also regularly used to remove the plaque (deposits of fatty tissue) inside blood vessels in cardiovascular disease, which is the number one cause of death in the USA and most of the developed and developing countries. Rotational atherectomy is a catheter-based procedure using a high-speed metal-bond diamond grinding wheel to pulverize the plaque and restore the blood flow in the vessel and health of the patient.

Two types of rotational atherectomy grinding wheels are shown in Fig. 14.18. Figure 14.18a shows a diamond wheel, typically rotating at 140,000–210,000 rpm, grinding the plaque. Another design, as shown in Fig. 14.18b, uses the orbiting diamond wheel (typically rotating at 60,000–120,000 rpm) and its combination of rotational and orbiting motion to grind the plaque. Details of the design and operation of orbital atherectomy grinding are illustrated in Fig. 14.19. The guidewire (0.36 mm in diameter) is stationary. Outside the guidewire is the rotational driveshaft, which is fabricated by spiraling six 0.15 mm diameter wires. Outside the driveshaft is the stationary catheter (0.18 mm OD and 0.16 mm wall thickness). Saline is slowly fed through the catheter (about 45 ml per min flow rate) during the atherectomy procedure to serve as the lubricant and fluid for the hydrodynamic bearing between the high-speed rotary driveshaft and stationary guidewire and catheter. The saline also reduces the heat, which is critical because the procedure is performed inside the blood vessel and excess heat can cause blood coagulation. The grinding wheel, also called crown, is a titanium rod, 1.98 mm in diameter, with bevel on both ends. The diamond has 0.07 mm average size. The shaft is placed eccentrically 0.5 mm from

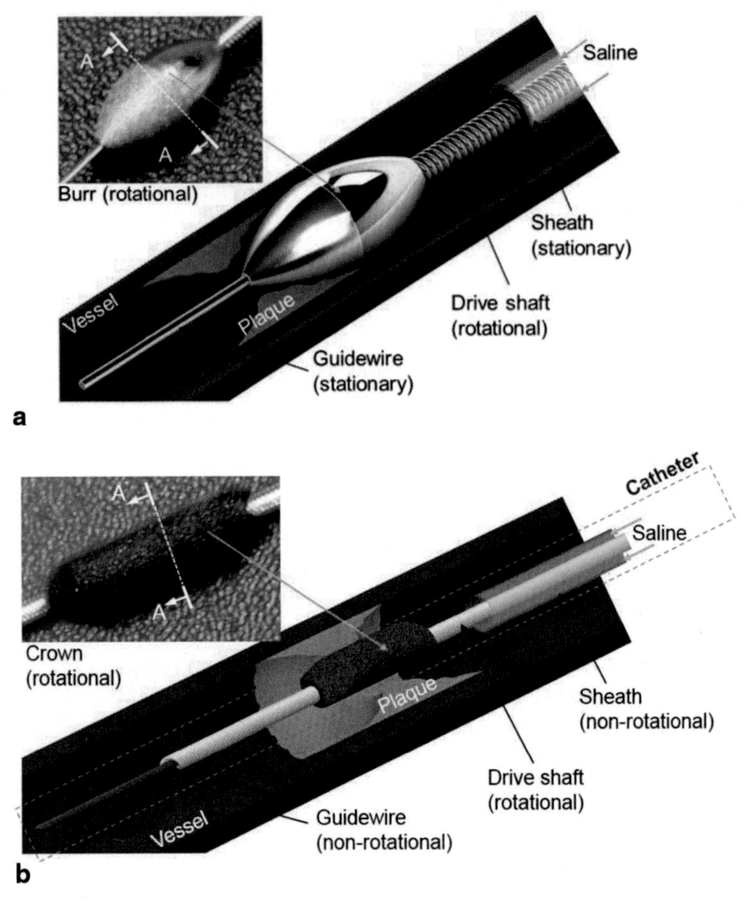

Fig. 14.18 Metal-bond diamond grinding wheels for plaque grinding in two types of the **a** rotational and **b** orbital atherectomy

the center of the rod. Such eccentricity generates the combination of the rotational and orbital motion of the grinding wheel. For example, at 90,000 rpm (1500 Hz) rotational speed, the grinding wheel is orbiting at about 38 Hz in an experiment using saline as the replacement of blood.

The grinding process breaks up the plaque into very small particles (typically smaller than the 6–8 μm red blood cells), which can pass harmlessly through the circulatory system and will eventually be absorbed by the body. This plaque grinding process, first introduced in 1993, is common in treating four types of plaques: (1) hardened plaque due to calcium deposits, (2) plaque located at branch points, (3) the recurrence of plaque in a previously placed stent, and (4) plaque near the joints of peripheral artery. These are procedures in which it is difficult to use stents to maintain the blood flow in the artery. Atherectomy is a plaque cutting process. It is a good demonstration of the biomedical machining in health care.

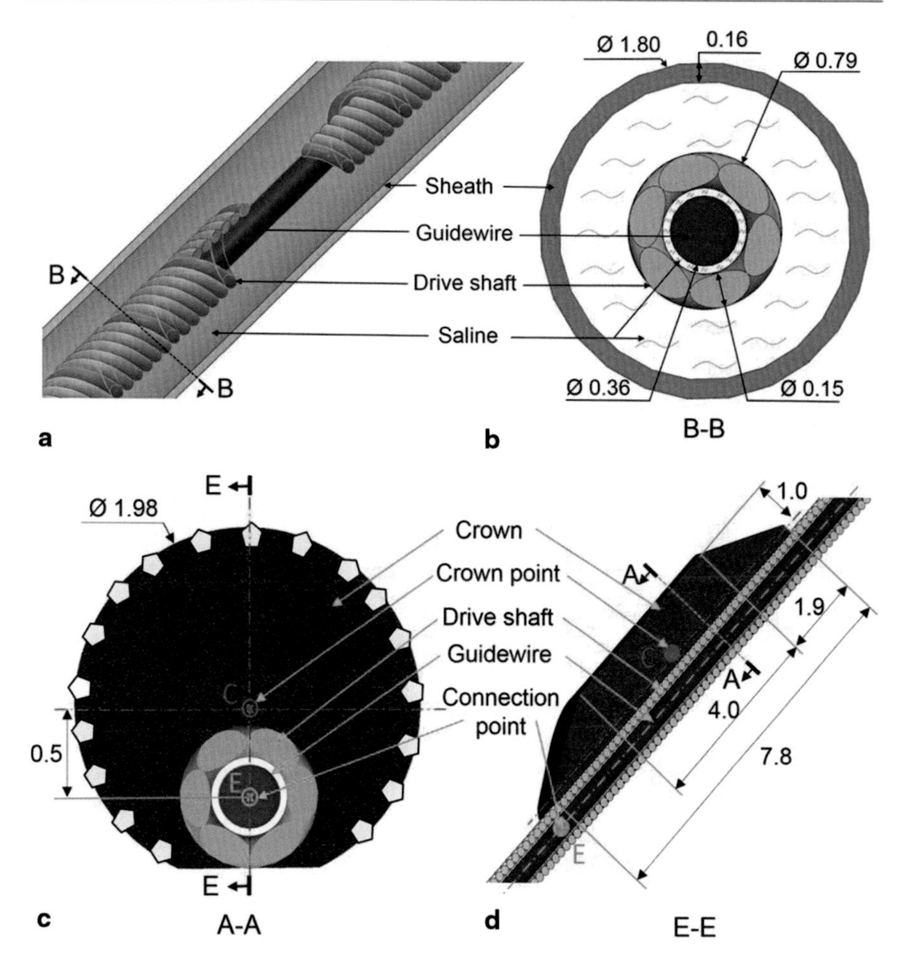

Fig. 14.19 The catheter and grinding wheel geometry in orbital atherectomy. **a** The guidewire, driveshaft, and sheath, **b** cross-section B-B, **c** crown cross-section A-A, and **d** cross-section E-E. (unit: mm)

14.7 Electrosurgical Processes and Thermal Dose Model

In surgery, surgeons need to cut tissue and coagulate blood using heat to form clots for hemostasis. In the past, a heated, sharp blade was used to cut and coagulate tissue at the same time in surgery. This procedure is performed by electrosurgical devices today. The first electrosurgical device was developed by Harvey Cushing, a surgeon, and William Bovie, a physicist, in 1926. The fundamental principles used in electrosurgery are similar to EDM and resistance heating.

Because human nerve and muscle stimulations cease at frequencies over 100 kHz, the electrical energy of alternating current of higher frequency can be used safely to generate the arc used for cutting tissues and coagulation of blood by heat. Without coagulation, blood would spread around the area of the cut and block the view of the operating surgeon. Heat has a significant side effect of damaging adjacent tissue, particularly the nerve or neurovascular bundle (NVB). Cells are sensitive to temperature. The threshold temperature for cellular thermal damage starts around 43 °C. In some surgical operations, for example, prostatectomy for men, hysterectomy for women, and neurosurgery in general, heat used for hemostasis may damage nearby nerve tissue. The problem is analogous to that of minimizing workpiece thermal damage, such as the white layer in hard turning and camshaft grinding surface thermal cracks and material phase transformation, in machining.

A thermal dose model relates the tissue temperature to the time of exposure until tissue death. Thermal dose commonly correlates with a corresponding dose at 43 °C. For example, tissue death is reached when the equivalent thermal dose at 43 °C reaches an application time of 20 min. The definition of a tissue thermal dose, also known as cumulative equivalent minutes (CEM) at 43 °C, denoted as CEM_{43}, as a function of the treatment temperature $T(t)$ and time t, is:

$$CEM_{43} = \sum_{t=0}^{t_e} R^{43-T} \Delta t \tag{14.10}$$

where t is the time (in seconds) of heating from 0 to the end t_e, T is the temperature of the tissue, and R is an empirical constant. Typically, $R=0$ for $T<39$ °C, $R=0.25$ for $39<T<43$ °C, and $R=0.5$ for $T>43$ °C.

For $CEM_{43}=20$ min (1200 s) at 43 °C, when the temperature is raised to 45 °C, it takes only 5 min to reach CEM_{43}. At 50 °C, the time for thermal injury is reduced to only 0.16 min (9 s). There is a critical need to control the temperature and develop knowledge and surgical thermal management devices for biomedical machining processes that use or generate heat.

14.8 Concluding Remarks

Biomedical machining is closely related to the development of advanced medical devices. Understanding the basic machining principles is the foundation for innovations and advancements in biomedical machining procedures, which can be in the operating room or in our daily living (such as needle insertion to soft tissue). Biomedical machining has broad applications and will continue to evolve as new needs in health care arise.

Homework

1. What is the inclination angle at the point A for sawing with bevel angle ξ and cutting velocity components v_s and v_w?

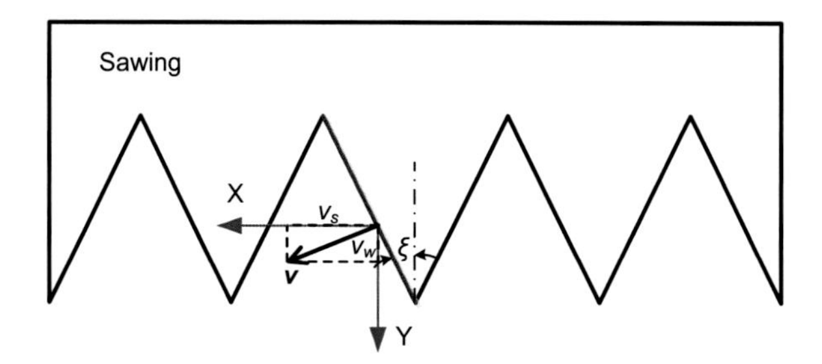

2. Draw the inclination angle and rake angle of a double-plane bevel needle with 10° bevel angle.

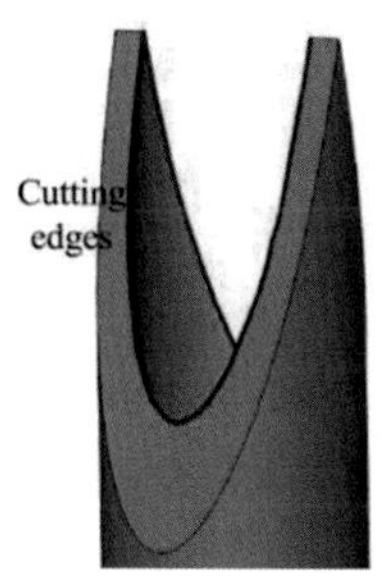

3. For an 18-gauge extra thin wall needle with $\xi=12°$, $\varphi=18°$, and $\beta=60°$, what is the q for grinding the needle with lancet point?
4. What is the needle with lancet point grinding setup parameter q for needles B and C in Fig. 14.14b. Note: Needle A is the example in Section 14.4.

References

Altintas Y (2012) Manufacturing automation, metal cutting mechanics, machine tool vibrations, and CNC design, 2nd edn. Cambridge University Press, Cambridge
Childs THC, Maekawa K, Obikawa T, Yamane Y (2000) Metal machining, theory and applications. Wiley, Hoboken
Cook NH (1966) Manufacturing analysis. Addison-Wesley, New York
DeVries WR (1991) Analysis of material removal processes. Springer, New York
Dudzinski D, Molinari A, Schulz H (2002) Metal cutting and high speed machining. Kluwer Academic, New York
Grzesik W (2008) Advanced machining processes of metallic materials, theory, modelling and applications. Elsevier, San Diego
HMT Limited (1999) Mechatronics and machine tools. McGraw-Hill, New York
Knight WA, Boothroyd G (2005) Fundamentals of machining and machine tools, 3rd edn. CRC Press
Krar SF, Rapisarda M, Check AF (1996) Machine tool and manufacturing technology. Delmar Publishers, New York
Marinescu ID, Ispas C, Boboc D (2002) Handbook of machine tool analysis. Marcel Dekker, Inc., New York
McCarthy WJ, Repp VE (1978) Machine tool technology. McKnight Publishing Company
Oxley PLB (1989) Mechanics of machining, an analytical approach to assessing machinability. Ellis Horwood
Reshetov DN, Portman VT (1988) Accuracy of machine tools. ASME Press, New York
Shaw MC (1984) Metal cutting principles. Clarendon Press, Oxford
Shaw MC (1996) Principles of abrasive processing. Clarendon Press, Oxford
Smith GT (1993a) CNC machining technology 1, design, development and CIM strategies. Springer, London
Smith GT (1993b) CNC machining technology 2, cutting, fluids and workholding technologies. Springer, London
Smith GT (1993c) CNC machining technology 3, design, part programming techniques. Springer, London
Stephenson DA, Agapiou JS (2016) Metal cutting theory and practice, 3rd edn. CRC Press
Tlusty J (1999) Manufacturing processes and equipment. Prentice Hall
Ulsoy AG, DeVries WR (1989) Microcomputer applications in manufacturing. Wiley, New York
Weck M (1984a) Handbook of machine tools, volume 1, types of machines, forms of construction and applications. Wiley, New York
Weck M (1984b) Handbook of machine tools, volume 2, construction and mathematical analysis. Wiley, New York

S. Y. Liang, A. J. Shih, *Analysis of Machining and Machine Tools*,
DOI 10.1007/978-1-4899-7645-1

Weck M (1984c) Handbook of machine tools, volume 3, automation and controls. Wiley, New York
Weck M (1984d) Handbook of machine tools, volume 4, metrological analysis and performance tests. Wiley, New York
Welcourn DB, Smith JD (1970) Machine-tool dynamics, an introduction. Cambridge University Press, Cambridge
Yoram Koren (1983) Computer control of manufacturing systems. McGraw-Hill, New York

Printed in the United States
By Bookmasters